The Search for Extraterrestrials

Intercepting Alien Signals

Monte Ross

The Search for Extraterrestrials

Intercepting Alien Signals

Published in association with
Praxis Publishing
Chichester, UK

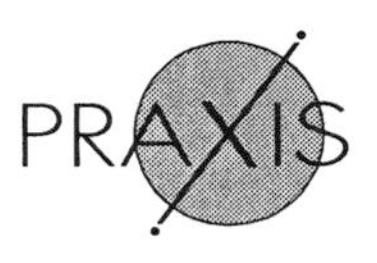

Mr Monte Ross
Chairman
Laser Space Signal Observatory
Olivette
Missouri
USA

SPRINGER–PRAXIS BOOKS IN POPULAR ASTRONOMY
SUBJECT *ADVISORY EDITOR*: John Mason, M.B.E., B.Sc., M.Sc., Ph.D.

ISBN 978-0-387-73453-8 Springer Berlin Heidelberg New York

Springer is a part of Springer Science + Business Media (*springer.com*)

Library of Congress Control Number: 2009925768

Cover design: Jim Wilkie
Project Copy Editor: David M. Harland
Typesetting: BookEns Ltd, Royston, Herts., UK

Printed in Germany on acid-free paper

To Harriet,
our children
Ryia, Diane & Ethan,
and our grandchildren
Jake, April, Gabriella & Julia.

Contents

Illustrations

Author's preface

In writing this book, I set out to educate, entertain, and stimulate public support for intelligent efforts by humankind to search for alien signals from space. As the book explains the primary facts and concepts in layman's terms, the reader will need only elementary mathematics to grasp the major issues. Whilst some oversimplification has been inevitable, the coverage is scientifically accurate. I hope the reasoning will resonate with the public. Once readers have grasped the fundamentals, I believe they will agree with me that the SETI effort to-date has been extremely limited in scope; in particular, it has not really explored the electromagnetic spectrum outside the very narrow radio-frequency region. In fact, not even 1 per cent of the spectrum has been explored!

The people involved in SETI are trying to achieve a technically challenging task in a manner consistent with the laws of physics. It has absolutely nothing to do with the "I was kidnapped by aliens" reporting of the tabloid press. Unfortunately, in the public mind UFOs and alien signals tend to get lumped together, with the result that SETI work is often portrayed as being on the fringe of science. Certainly, in the past, many engineering and scientific academics risked SETI tarnishing their careers. The low probability of success is a barrier to academics interested in furthering their careers. Some members of the public may believe a massive effort to detect alien signals is underway, but this is far from being the case. In fact, no US government funding currently exists for SETI. The limited efforts that are in progress rely on charitable funding. SETI has been mainly the work of academics well versed in a particular discipline. However, as I show in the text, the detection of extraterrestrial signals is a systems engineering problem whose solution – constrained by limited resources – both minimizes the chance of failure and maximizes the chance of success. SETI efforts to-date have been marked by basic deficiencies which make it unlikely they will succeed. Section 4 of this book offers some suggestions for how a systems engineering approach might be developed.

There has been a great deal of speculation in the literature about why we have had no success, with the inference being that this is because there is nobody to signal to us. This is known as the Fermi Paradox, but it is far too early to conclude that we are alone. Little has been done, especially at laser wavelengths. This book explains what has been done, what is being done, what can be done, and what may have to be done before we can in good conscience acknowledge the Fermi Paradox.

A related effort is that underway to identify Earth-like planets orbiting other stars. This was initially done using terrestrial telescopes, but recently satellites have joined the hunt. After years of frustration, in excess of 300 extrasolar planets have been detected in the last decade, in some cases with a single star having several planets. The early methods were best suited to finding gas giants even larger than Jupiter, but the latest systems are designed to find Earth-like planets. The latest addition to this effort is the Kepler spacecraft launched by NASA in March 2009. When it does find a planet with a mass comparable to Earth that is orbiting a solar-type star at just the right distance for the surface to be conducive to life, NASA will certainly hail it as a major discovery. And indeed it will be, but a dozen or so further requirements must be satisfied before a star system can produce an intelligence capable of attempting to communicate across interstellar distances.

This book seeks to inform the public of what has been done, what is being done, what can be done and what needs to be done, much of it at little cost and potentially great benefit to mankind. The obstacles to success are discussed in detail. There is a long way to go. SETI is limited to, at best, several million dollars per year globally. If we are serious, then we should do better than this. With public support, money can be forthcoming or the challenge may be taken up by a wealthy foundation.

SETI has been of interest to me personally since 1965, soon after the birth of the laser. In 1961 Robert Schwartz and Charles Townes wrote a scientific paper which explained the potential for lasers in extraterrestrial communications. However, little was done while attempts to detect signals concentrated on radio frequencies. Interest in optical SETI was kept alive by a few people, notably myself, Stuart Kingsley and Ragbir Bhathal, until the lack of success at radio frequencies forced reconsideration. I pointed out in 1965 that laser signals would be best sent by short pulses, since this would allow a modest transmitter to readily overcome the brightness of the host star. Today, there is a serious effort at Harvard devoted to pulsed laser signals, but all the efforts at all wavelengths are still quite limited in relation to what can be done. Thus, this book is in four sections. The first section discusses the likelihood of intelligent life, taking into account the many constraints which could conceivably mean that we are the only intelligence in the galaxy. The second section discusses the basics of space communication and the technical approaches and issues. The third section describes the SETI efforts that have been attempted and are currently underway. The final section explains how we might proceed further. The bibliography should assist anyone seeking a guide to the literature on the subject.

In my career, lasers have gone from being a new subject of interest, to finding use in a wide range of applications. I addressed laser communication in my 1965 book *Laser Receivers*, and spent my career helping to develop laser communication. It is gratifying that the advances in power and efficiency of lasers have facilitated such varied applications, and in my role as technical editor of a *Laser Applications* series by the Academic Press I kept up to date with the myriad uses of lasers in industry, defense, medicine and entertainment. I participated in these advances, being the first to patent a semiconductor diode pumped crystal laser that is ubiquitous today from a green laser pointer for public speakers, to a satellite-to-satellite communication link. As Director of Laser Space Systems for McDonnell Douglas for many years, and in working with both NASA and the United States Air Force, I was intimately involved with the development of space communication. In 1975 the IEEE gave me a Fellow Award, as did McDonnell Douglas in 1985, both for "Leadership and Contributions to laser communications". I believe there are excellent reasons for lasers being the best choice for SETI. I leave it to the reader to form a judgement on which part of the spectrum we are more likely to intercept a signal from space: radio-frequency, optical, infrared, ultraviolet, or something else. I discuss a variety of considerations, including information that makes it clear how difficult it is for interception to occur. The material on communication is based explicitly on my background and expertise. That on the likelihood of a planet giving rise to intelligence derives from the rapidly increasing body of knowledge of planetary science. Overall, I hope that this book provides a firm base for those who might wish to further the search for alien signals.

Finally, it should be borne in mind that signals may be being sent to us right now, but we will not know it unless we are looking in the right part of the spectrum at the right time, with the right equipment pointing in the right direction. Success may be tomorrow, or a millennium away.

Monte Ross
July 2009

Acknowledgments

I must thank Clive Horwood of Praxis for accepting this book for publication, John Mason for suggesting how it should be structured, and David M. Harland for editing the manuscript. I am also grateful to Miriam Pierro-Ross and Mark Peterson, who helped me to navigate the computer world. It would be difficult to thank each and every person whose work I have used in compiling the facts essential to this book, but I have mentioned in the text those individuals whose work had most impact and I have included an extensive bibliography. I must also thank Robert Olshan and Stephen Webb for their helpful suggestions. The illustrations are from varied sources. Some are off the Internet, but others were created specifically for this book. If I have not assigned a credit, and the owner notifies me via the publisher, I will be glad to rectify that omission in a future edition.

Part 1: The likelihood of extraterrestrial intelligence

1 Vast distances and long travel times

It has been said that the discovery of an extraterrestrial intelligence will be the most important event in mankind's history. For millennia, humans have been looking at the stars at night and wondering whether we are alone in the universe. Only with the advent of large-dish radio-frequency antennas and ultra-sensitive receivers in the late-twentieth century did it become possible to attempt a search for extraterrestrial intelligence (SETI).

The search at radio frequencies continues and has even involved the public (see SETI@home) by allowing home PCs to analyze some of the received noise. With so much data collected, it becomes easier to examine if pieces of the data are divided up and dispersed to many individual computers. A home PC can analyze the data at a time it is otherwise idle (see Chapter 8). The fact that tens of thousands of people signed up to participate illustrates the strong public interest in SETI. Whilst a very successful promotion, it has had no success in finding an extraterrestrial signal.

On the other hand, look at what we have accomplished in less than 200 years: we have progressed from essentially being limited to communicating within earshot or by messengers traveling on foot or riding horses, to communicating at the speed of light with space probes millions of kilometers away. This fantastic accomplishment illustrates the exponential growth of our technology. In this context, several decades spent on SETI is a mere drop in the bucket of time. The disappointment of SETI to-date is, I believe, due to the overoptimistic expectation of there being an advanced intelligence in our immediate neighborhood. Less than 100 years ago it was widely believed that there might be beings on Mars or Venus, the nearest planets to us. We now know this is not so. Indeed, we have come to realise that whilst intelligent life on planets orbiting other stars is feasible, its development is dependent on a number of conditions that may not occur in combination very often.

In spite of there being several hundred billion stars in our Milky Way galaxy, the likelihood of an intelligent society sending signals our way is thought to be low. The recent discovery of over 300 planets orbiting relatively nearby stars lends hope that there are many planets that can sustain life, some of which will develop intelligence that is willing to communicate. But the equation developed by Frank Drake in 1960 (Chapter 6), the hypothesis advocated by Peter Ward and Donald E. Brownlee in their book *Rare Earth: Why Complex Life is Uncommon in the Universe*, published in 2000 (Chapter 3), and the study by Stephen Webb using the Sieve of Eratosthenes in his book *If the Universe is Teeming with Aliens... Where is Everybody*, published in 2002 (Chapter 6), all highlight the many probabilities in play. Depending on how optimistic one is in assigning probabilities to each factor, one can reach either very low probabilities or much better odds. A probability of one in a million would still mean 400,000 stars in our galaxy have intelligent life – and there are hundreds of billions of galaxies. So where are they? Either intelligence is scarcer, or we have not been looking in the right places using the right instruments at the right time.

The failure of SETI to-date raises the intriguing question of whether our search at radio frequencies was naive, since no intelligent society would use radio frequencies to transmit over distances of hundreds of light-years if other wavelengths were more useful. Is a technology which we ourselves have only recently acquired likely to be favored by a far more advanced society? In fact, a good argument can be made that radio frequencies are an unlikely choice for an advanced society, and that if we must select just one part of the electromagnetic spectrum to monitor then visible, infrared or ultraviolet offer better prospects for SETI. In essence, the case against radio is that it is a high-powered transmission whose wide beam washes over many stars. In contrast, lasers in the visible, infrared or ultraviolet require less power and the energy is aimed towards a particular star system. A civilization seeking to establish contact with any intelligences around stars in its neighborhood might aim such a laser at a star which shows characteristics likely to support life. As so few star systems have such characteristics, we would probably be included in a targeted search by a nearby civilization. If we were fortunate, we might spot such a laser probing for a response from any life in our system. Although many papers have been written showing why and how laser signals could be present, early studies by radio-frequency engineers compared continuous-wave laser signals with continuous-wave radio frequencies and drew conclusions that may not actually be correct. It was clear from the physics and from the noise and background light that the most efficient modulation method at optical wavelengths was high-peak-power short-pulse low-duty-cycle pulses. The term short-pulse low-duty-cycle refers to the fact that the signal is not continuous, but is active only for a small fraction of the time. For example, the transmitted pulse may be on for one nanosecond, and the pulse rate may be once per millisecond. As the duty cycle is the pulse width multiplied by the pulse rate, we have 1 nanosecond multiplied by 1,000 pulses per second for a duty cycle of one part in a million. This means that the system is transmitting one-millionth of the time. Thus the peak power can be 1,000,000 times the average power, or the continuous power in this example. Other issues in determining the best choice for such communication are discussed in later chapters.

In retrospect, it is evident that SETI began searching at radio frequencies because extraterrestrial intelligence was initially believed to be plentiful and we had systems for receiving weak radio signals from probes operating in deep space, whereas laser technology was not at the same level of development.

The likelihood of radio frequencies being used in lieu of lasers is diminished if nearby star systems are not transmitting. This is due to the much larger antennas that would be required at the receiver site to receive signals from much greater

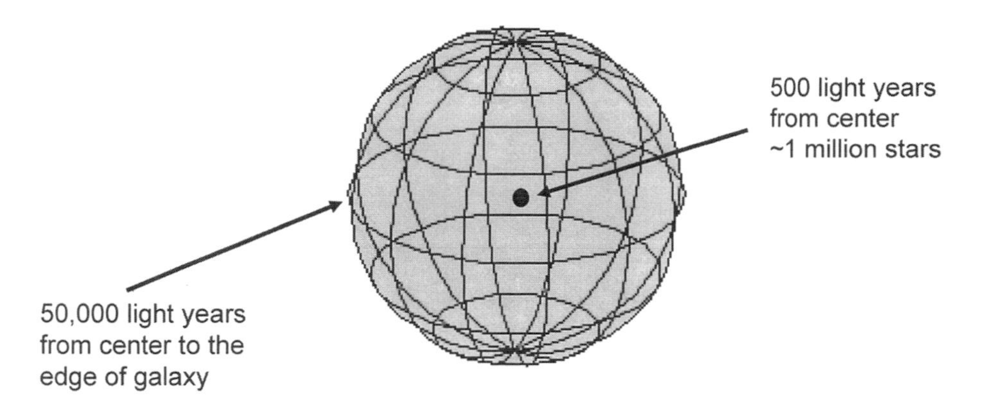

Figure 1.1. The vastness of our galaxy.

distances. The received power is proportional to the area of the antenna. A light-year is 9.46×10^{12} kilometers, and stars are many light-years apart. Owing to the inverse square law in which the area irradiated increases by the square of the distance, there is a factor of 400 difference in the signal power lost in space between a source that lies 10 light-years away and one 200 light-years away. If the same transmitter is used, the area of the receiving antenna must be increased by a factor of 400 in order to detect a source 200 light-years away compared to 10 light-years away (i.e. 20×20). This may well be impracticable. And this is only one argument against using radio frequencies for interstellar communication. It is more likely that the stars will be far away because of geometry. That is, imagine the Sun to be located at the center of a sphere in which the other stars are assumed to be more or less equally distributed (Figure 1.1), then the fact that volume is a function of the cube of distance means that there will be 8 times more star systems within a radius of 100 light-years from the Sun than a radius of 50 light-years, and 64 times more within 200 light-years. It is therefore 512 times more likely that an intelligent society may be sending us signals if we look to a distance of 400 light-years rather than a distance of 50 light-years. Figure 1.2 shows that there are approximately 1 million stars similar to the Sun within a radius of 1,000 light-years. However, as constraints are applied and more is learned about potential star systems, the probability of there being anyone signaling to us continues to decline.

How far are the stars and how do we know?

One question that is often asked is how we know stellar distances. One of the major ways is to use the parallax effect. As shown in Figure 1.3, parallax measures the angle to a point from two vantage points. The distance to that point can be calculated by applying simple trigonometry to the angular measurements. The distance between the vantage points is the baseline, and the longer the baseline

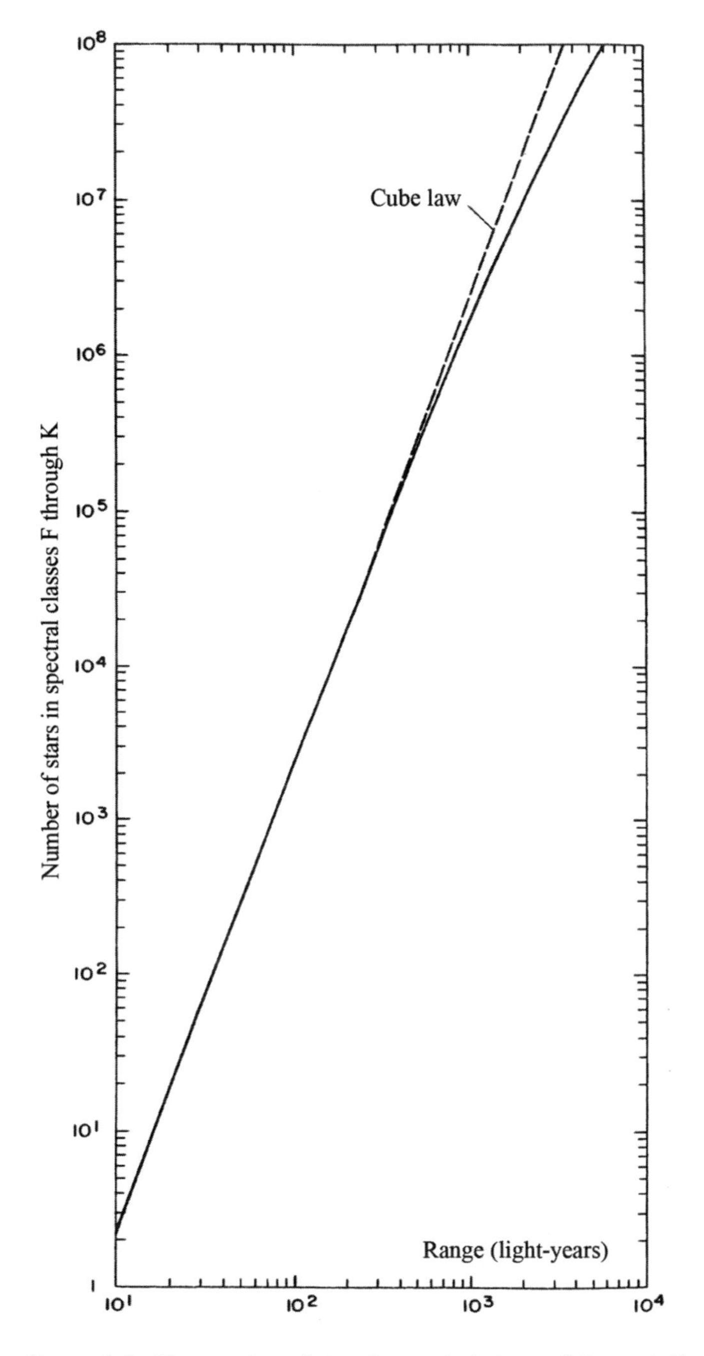

Figure 1.2. The number of stars in spectral classes F through K.

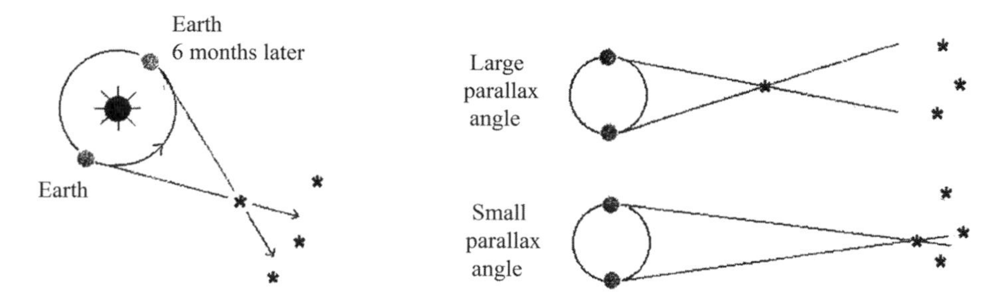

Figure 1.3. The parallax effect (left) and parallax change with distance (right).

the more accurate the distance measurement. The longest baseline available to a terrestrial observer is the diameter of Earth's orbit around the Sun. A star observed at suitable times 6 months apart will appear in a different position on the sky as the angle of viewing changes slightly. The closer the star, the greater its parallax and the more it will be displaced relative to the background of more distant stars. However, even for nearby stars the effect is small, and highly accurate measurements are required to obtain results with high confidence. The annual parallax is defined as the angle subtended at a star by the mean radius of Earth's orbit of the Sun. A 'parsec' is 3.26 light-years, and is based on the distance from Earth at which the annular parallax is one second of arc. The angles are very small because the distance across Earth's orbit of the Sun is extremely small in comparison to the distances of the stars. Indeed, the nearest star, Proxima Centauri, lies 4.3 light-years away and has a parallax of only 0.76 seconds of arc.

The accuracy of angular measurements made from Earth's surface is limited by distortions in the atmosphere. Telescopes in space therefore have an advantage. In 1989 the European Space Agency put a satellite named Hipparcos into orbit around Earth to employ the baseline of Earth's orbit around the Sun to accurately measure parallaxes for stars as far away as 1,600 light-years. There are methods which do not use geometric parallax and facilitate measurements at greater distances. These are more difficult to implement, but can yield reasonably accurate results. In 1997 NASA began a study of a Space Interferometry Mission (SIM). Progress was slow due to budget constraints. As currently envisaged, the renamed SIM Lite will be launched at some time between 2015 and 2020 and be put into solar orbit trailing Earth. It will have a number of goals, including searching for terrestrial planets in nearby star systems. Optical interferometry will enable the positions of stars on the sky to be measured to within 4 millionths of a second of arc. This will facilitate measuring distances as far away as 25 parsecs to an accuracy of 10 per cent, which is many times better than is possible from Earth's surface.

By a variety of techniques the parallax effect can provide acceptable results out to about 1,000 light-years, with the distances to the nearer stars being more accurate than those farther away. Such a volume of space includes a large number of stars. It can therefore be assumed that an advanced civilization will accurately know how far we are from them, and hence can calculate the transmitter power needed to reach us.

Of course, another issue is the time involved in communicating across interstellar distances, because an electromagnetic signal traveling at the speed of light takes one year to travel a light-year. A civilization might be willing to try to prompt a response from a nearby star system, but reject waiting hundreds of years for a response from a distant star. The volume of space within which communication is practicable might therefore be quite small.

2 Stars, their evolution and types

In the last few years we have been able to detect a number of extra-solar planetary systems, but we cannot tell much about them. Our knowledge will improve in the next decade or two, however. It is likely that an advanced extraterrestrial civilization will know which star systems in its neighborhood are good candidates to host intelligent life, and which are not. The primary selection criteria are the type of the star, which is related to the temperature of its surface, and the size and location of its planets. As we learn more about planets and their characteristics, we should be able to apply a variety of other constraints (Chapter 6). Once an advanced society has made such an analysis, the resulting list of nearby stellar systems likely to harbor life may well be very short.

To understand the search for intelligent extraterrestrial signals, it is necessary to consider the hundreds of billion stars in our galaxy which are possible hosts, and the means of transmission and reception of a signal over such large distances.

Consider the problem of a civilization which wishes to contact another intelligent society. How do they proceed? They appreciate that conditions for intelligent life are quite restrictive, but conclude that there are so many stars that perhaps all they need to do is to make a thorough search. But the galaxy is approximately 100,000 light-years across, and communication across that distance would be impracticable. It would be better if they were to find a society within about 500 light-years. Although small in relation to the galaxy as a whole, this volume is likely to include in excess of a million stars, which is a reasonable basis for applying the 'habitability' selection criteria.

To better understand the likelihood of advanced intelligence in our galaxy, it is worth reviewing the types and evolution of stars, and the chance of one possessing a planet with characteristics suitable for the development of an advanced intelligence. However, much of what we have inferred is based on the only intelligent life that we know of, namely ourselves and our environment in the solar system, and there is the possibility that we are in some way atypical. Nevertheless, with this caveat in mind it is possible to estimate the likelihood of other stars having planets that are in this sense 'right' for the development of advanced intelligence.

In what follows, we will examine the constraints imposed on stellar systems as suitable abodes of intelligent life. Some constraints seem certain, some seem likely, and others are simply possibilities about which cosmologists argue. As we discover more about stellar systems, the individual constraints may be tightened or loosened. In general, as we have learned more, the probability of there being

Distance (light-years)	Number of Stars
50,000	400×10^9
25,000	50×10^9
12,500	6.25×10^9
6,250	$\sim 8 \times 10^8$
3,125	$\sim 1 \times 10^8$
1,500	$\sim 1.2 \times 10^7$
750	$\sim 1.5 \times 10^6$
400	$\sim 2 \times 10^5$
200	$\sim 2.5 \times 10^4$
100	~3,000
50	~375
25	~47
12.5	~ 6

Figure 2.1. The number of stars within a given distance of the Sun.

another advanced society nearby has reduced. Indeed, if the constraints are applied harshly it becomes unlikely that there is another intelligent civilization anywhere near us.

In the ancient past, Earth was considered to lie at the center of the universe, with mankind being special. The work of Copernicus and Galileo in the sixteenth and early seventeenth centuries showed that the planets, including Earth, travel around the Sun. This weakened man's perception of being centrally located. The discovery that there are hundreds of billions of stars in the galaxy and hundreds of billions of galaxies provided a sense of immensity that reinforced man's insignificance. But the possibility that we are the only advanced civilization puts us center-stage again. To assess the chances of there being many societies out there, we need to know more about stars and planets. Figure 2.1 shows the number of stars within a given radius of us.

A galaxy such as ours comprises a spherical core and a disk that is rich in the gas and dust from which stellar systems are made. The interstellar medium is typically composed of 70 per cent hydrogen (by mass) with the remainder being helium and trace amounts of heavier elements which astronomers refer to as 'metals'. Some of the interstellar medium consists of denser clouds or nebulas. Much of the hydrogen in the denser nebulas is in its molecular form, so these are referred to as 'molecular clouds'. The largest molecular clouds can be as much as 100 light-years in diameter. If a cloud grows so massive that the gas pressure cannot support it, the cloud will undergo gravitational collapse. The mass at which a cloud will collapse is called the Jeans' mass. It depends on the temperature and density, but is typically thousands to tens of thousands of times the mass of the Sun. As the cloud is collapsing, it may be disrupted by one of several possible events. Perhaps two molecular clouds come into collision with each other. Perhaps a nearby supernova explosion sends a shock wave into the cloud. Perhaps two galaxies collide. By such means, clouds are broken into condensations known as Bok globules, with the smallest ones being the densest. As the process of collapse continues, dense knots become protostars and the release of gravitational energy causes them to shine. As the protostar draws in material from the surrounding cloud, the temperature of its core increases. When the pressure and temperature in the core achieve a certain value, nuclear fusion begins. Once all the available deuterium has been fused into helium-3, the protostar shrinks further until the temperature reaches 15 million degrees and allows hydrogen to fuse into helium, at which time radiation pressure halts the collapse and it becomes a stable star.

The onset of hydrogen 'burning' marks the initiation of a star's life on what is called the 'main sequence' of a relationship derived early in the twentieth century by Ejnar Hertzsprung and Henry Norris Russell. They plotted the absolute magnitudes of stars against their spectral types, observational parameters which equate to the intrinsic luminosity and surface temperature. The resulting diagram (Figure 2.2) shows a high correlation between luminosity and surface temperature among the average-size stars known as dwarfs, with hot blue stars being the most luminous and cool red stars being the least luminous. Running in a narrow band from the upper left to the lower right, this correlation defines the main sequence. Its importance is that all stars of a given mass will join the main sequence at a given position. But stars evolve and depart the main sequence. If a star becomes a giant or a supergiant, it will develop a relatively high luminosity for its surface temperature and therefore move above the main sequence. If a star becomes a white dwarf, its luminosity will be relatively low for its surface temperature, placing it below the main sequence. The stars that lie on the main sequence maintain a stable nuclear reaction, with only minor fluctuations in their luminosity. Once the hydrogen in its core is exhausted, a star will depart the main sequence. The more massive the star, the faster it burns its fuel and the shorter its life on the main sequence. If the development of intelligent life takes a long time, then it might be limited to low-mass stars. The actual ages of stars are known only approximately, but it is clear that whilst very massive stars can remain on the main sequence for only several million years, smaller ones should do so for 100 billion years. Since the universe is 13.7 billion years old, it is evident that many low-mass stars are still youthful. The Sun is believed to have condensed out of a nebula about 5 billion years ago and to be half way through its time on the main sequence.

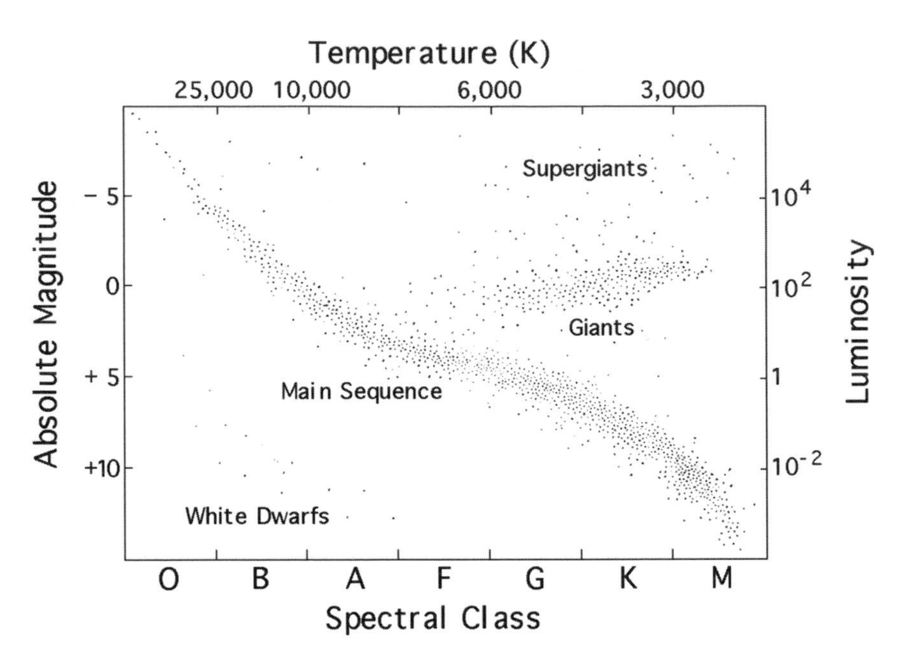

Figure 2.2. The Hertzsprung–Russell diagram showing phases of stellar evolution, in particular the main sequence.

At 1.99 × 10^{30} kg, the Sun is 333,000 times the mass of Earth. Astronomers find it convenient to express stellar masses in terms of the solar mass. The range of stellar masses is believed to result from variations in the star formation process. This theory suggests that low-mass stars form by the gravitational collapse of rotating clumps within molecular clouds. Specifically, the collapse of a rotating cloud of gas and dust produces an accretion disk through which matter is channeled 'down' onto the protostar at its center. For stars above 8 solar masses, however, the mechanism is not well understood. Massive stars emit vast amounts of radiation, and it was initially believed that the pressure of this radiation would be sufficient to halt the process of accretion, thereby inhibiting the formation of stars having masses exceeding several tens of solar masses, but the latest thinking is that high-mass stars do indeed form in a manner similar to that by which low-mass stars form. There appears to be evidence that at least some massive protostars are surrounded by accretion disks. One theory is that massive protostars draw in material from the entire parent molecular cloud, as opposed to just a small part of it. Another theory of the formation of massive stars is they are formed by the coalescence of stars of lesser mass. Although many stars are more massive than the Sun, most are less so. This is a key issue in estimating the prospects for the development of life, as the lower surface temperature of a smaller star sharply reduces the number of photons with sufficient energy for the process of photosynthesis. The color of a star defines its spectral class, and by assuming that it acts as a 'blackbody' and radiates its energy equally in all directions it is possible to calculate the temperature of its surface. The hottest stars have their peak wavelength located towards the ultraviolet end of the visible spectrum, but the coolest stars peak in the infrared. When astronomers in the early twentieth century proposed a series of stages through which a star was presumed to pass as it evolved, they introduced an alphabetical sequence. Although further study prompted them to revise this process, the alphabetical designations were retained and the ordering was changed. Hence we now have O-B-A-F-G-K-M, where O stars are blue, B stars are blue-white, A stars are white, F stars are white-yellow, G stars are yellow, K stars are orange, and M stars are red. Other letters were added later. For example, R, S and C are stars whose spectra show specific chemical elements, and L and T signify brown dwarfs. The spectral class is further refined by a numeral, with a low number indicating a higher temperature in that class. Hence, a G1 star will have a higher temperature than a G9. The surface temperatures of stars on the main sequence range from around 50,000K for an O3 star, down to about 2,000K for an M9 star. With a spectral class G2 and a surface temperature of ~5,700K, the Sun is a hot-yellow star.

In general, a star will spend 80 per cent of its life on the main sequence but, as we have noted, more massive stars do not last very long. If they do possess planets, these probably do not have time for intelligence to develop. Once the hydrogen in the core is consumed, the star will evolve away from the main sequence. What happens depends on its mass. For a star of up to several solar masses, hydrogen burning will continue in a shell that leaves behind a core of inert helium. In the process, the outer envelope is inflated to many times its original diameter and simultaneously cooled to displace the peak wavelength towards the red end of the visible spectrum, turning it into a red giant of spectral classes K or M. When more massive stars evolve off the main sequence they not only continue to burn hydrogen in a shell, their cores are hot enough to initiate helium fusion and this additional source of energy inflates the star into a red supergiant. Such stars may well end their lives as supernovas. Stars which have left the main sequence are rarely stable, and even if life developed while the star was on the main sequence, this will probably be extinguished by its subsequent evolution. Certainly when the Sun departs the main sequence it will swallow up the inner planets.

Dwarfs of class K or M have surface temperatures of between 4,900K and 2,000K. They will last a very long time, longer indeed than the universe is old. This explains why they are so numerous. It may be that many red dwarfs possess planets, but the low temperature has its peak emission in the red and infrared, with the result that most of the photons are weak, possibly too weak to drive photosynthesis. If a planet is located sufficiently close to the star for its surface to be warm enough for life, the gravitational gradient will cause the planet to become tidally locked and maintain one hemisphere facing the star. (The change in rotation rate necessary to tidally lock a body B to a larger body A as B orbits A results from the torque applied by A's gravity on the bulges it has induced on B as a result of tidal forces. It is this process that causes the Moon always to face the

Spectral class data

Class	Surface temperature (K)	Color	Main sequence (Myr)
O	25,000–50,000	Blue	1-10
B	10,000–25,000	Blue-white	11–400
A	7,500–10,000	White	400–3,000
F	6,000–7,500	White-yellow	3,000–7,000
G	4,900–6,000	Yellow	7,000–15,000
K	3,500–4,900	Orange	~17,000
M	2,000–3,500	Red	~56,000

same hemisphere to Earth.) Thus, if planets around red dwarfs are a similar distance from their primaries as Earth is from the Sun they might lack sufficient energy for the development of life, and if they are close enough to obtain the necessary energy they will be tidally locked and it is not known whether life can survive on a tidally locked planet: if there is an atmosphere, the resulting intense storms will not be conducive to life. The conditions for life are better in spectral classes F and G. However, whilst this is consistent with the fact that we live in a system with a G star, we must recognize that our analysis is biased towards life as we know it.

As noted, most stars are class M red dwarfs. Figure 2.3 shows that of the 161 stars within 26 light-years of the Sun, 113 are red dwarfs, which is in excess of 70 per cent. Although this proportion may vary throughout the galaxy, it illustrates the fact that most stars are cooler than the Sun. How does this affect the prospects for life? Figure 2.4 illustrates the peak wavelength and intensity of a star's output as a function of wavelength. At lower temperatures the peak shifts towards the infrared. The peak wavelength for a 4,000K star is 724 nanometers, just inside the visible range. For a 3,000K star not only is the peak displaced into the infrared, at 966 nanometers, the intensity of the peak is significantly different. The intensity of the peak for a 6,000K star is over five times that of a 4,000K star. This represents a severe obstacle to the development of intelligent life in a red dwarf system. Perhaps the most fundamental issue is the paucity of energy in the visible and ultraviolet to drive photosynthesis. As Albert Einstein discovered, the photoelectric effect is not simply a function of the number of photons, it requires the photons to be of sufficiently short wavelengths to overcome the work function of an electron in an atom and yield a photoelectron. In a similar fashion, photosynthesis requires energetic photons. In the following chapters we will explore a number of factors that may preclude the development of intelligent life on most planets.

A few words should address the well-known star constellations, and point out just how distant the stars in a constellation are from each other. Astrological inventions such as the 'Big Dipper' represent patterns drawn in the sky by our ancestors, but in reality the stars of a constellation are not only unrelated to each

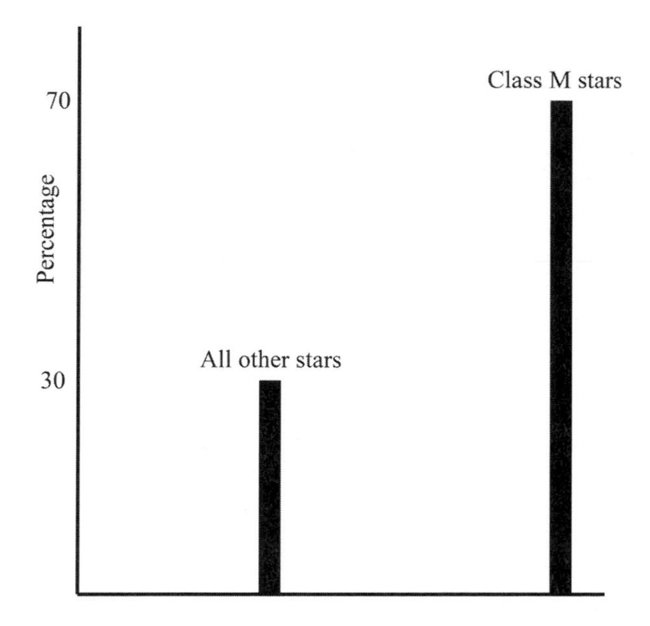

Figure 2.3. The majority of stars are of spectral class M.

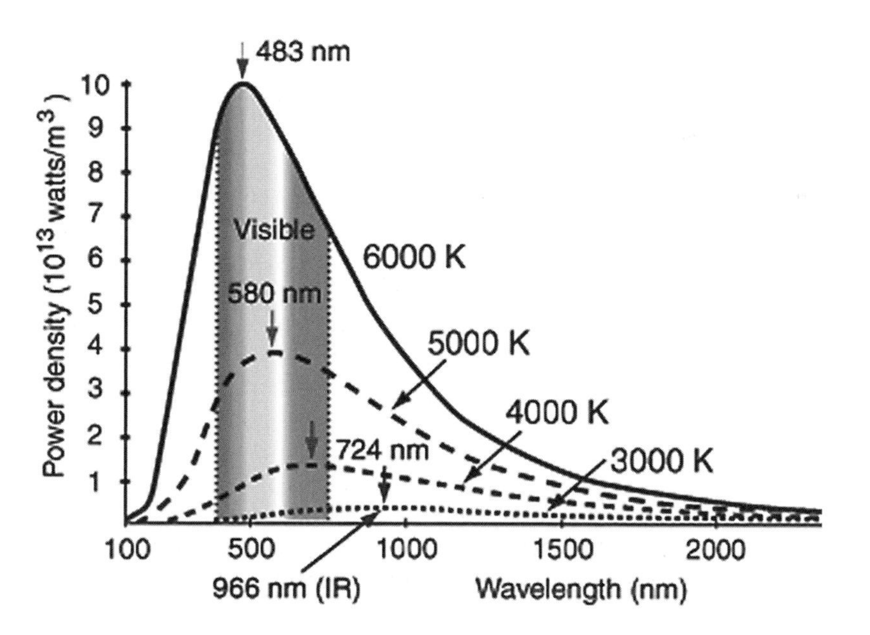

Figure 2.4. The emission peak of blackbody radiation as a function of wavelength.

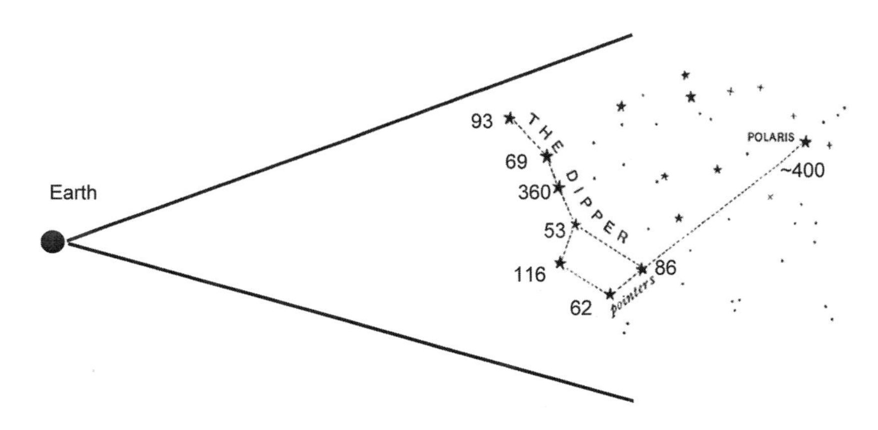

Figure 2.5. The distances to the stars in the 'Big Dipper' measured in light-years.

other they are also at distances ranging between 53 and 360 light-years (Figure 2.5). For SETI therefore, the constellations have no intrinsic significance.

Threats to life

At this point, we should outline how dangerous the universe is. Supernovas are stars that explode and not only issue ionizing radiation but also send shock waves through the interstellar medium. They shine for a short time, often only a few days and rarely more than several weeks, with an intensity billions of times that of the Sun. Their expanding remnants remain visible for a long time. Recent studies suggest that for a galaxy like ours a supernova will occur every 50 years on average. If a supernova were to occur close to a stellar system that hosted advanced life, it could essentially sterilize that system. Fortunately, where we are in the galaxy, supernovas should occur no more frequently than every 200 to 300 million years.

Figure 2.6 shows the 'great extinctions' of life on Earth, known as the Ordovician, Devonian, Permian, Triassic–Jurassic and Cretaceous–Tertiary. The worst is thought to have been the Permian, where 96 per cent of all marine species and 70 per cent of terrestrial vertebrate species died off. The second worst, the Ordovician, could well have been caused by a supernova 10,000 light-years away that irradiated Earth with 50 times the solar energy flux, sufficient to destroy the chemistry of the atmosphere and enable solar ultraviolet to reach the surface. Far worse would be a supernova at 50 light-years. The atmosphere would suffer 300 times the amount of ionization that it receives over an entire year from cosmic rays. It would ionize the nitrogen in the atmosphere, which would react with oxygen to produce chemicals that would reduce the ozone layer by about 95 per cent and leave the surface exposed to ultraviolet at an intensity four orders of magnitude greater than normal. Lasting for 2 years, this would probably sterilize the planet. Astronomer Ray Norris of the CSIRO Australia Telescope National

Ordovician
~488-439 million years ago

Devonian
~360 million years ago

Permian
~251 million years ago

Triassic-Jurassic
~200 million years ago

KT (Cretaceous-Tertiary)
~65.6 million years ago

Figure 2.6. The five 'great extinctions' of Earth history.

Facility estimated that a supernova should occur within 50 light-years once every 5 million years. In fact, they occur rather less frequently. Nevertheless, a nearby supernova would pose a serious threat to intelligent life.

Gamma-ray bursters, the most powerful phenomenon in the universe, also pose a threat to life. All those seen to-date occurred in other galaxies. They appear to occur at a rate of one per day on average. Studies suggest that in a galaxy such as ours, a gamma-ray burster will occur once every 100 million years. They are more powerful than supernovas, but when one flares it lasts less than a minute. We have observed slowly fading X-ray, optical and radio afterglows. Although intense, the gamma-ray flash is so brief that only one hemisphere of Earth would be irradiated, allowing the possibility of survival for those living on the other side of the planet. Nevertheless, world-wide damage would result. The ultraviolet reaching Earth would be over 50 times greater than normal. It would dissociate molecules in the stratosphere, causing the creation of nitrous oxide and other chemicals that would destroy the ozone layer and enshroud the planet in a brown smog. The ensuing global cooling could prompt an ice age. The significance of gamma-ray bursters for SETI is that if such an outburst sterilizes a large fraction of a galaxy, perhaps there is no-one left for us to eavesdrop on.

Magnetars are neutron stars which emit X-rays, gamma rays and charged-particle radiation. There are none in our part of the galaxy, but in the core a vast number of stars are closely packed and neutron stars are common. Intelligent life in the galactic core must therefore be unlikely.

On a more local scale, there is always the threat of a planet being struck by either an asteroid or a comet. Most of the asteroids in the solar system are confined to a belt located between the orbits of Mars and Jupiter, but some are in elliptical orbits which cross those of the inner planets. The impact 65 million years ago which wiped out half of all species, including the dinosaurs, is believed to have been an asteroid strike. We have a great deal to learn about other star systems, but if asteroid belts are common then they could pose a serious threat to the development of intelligent life there.

Life might wipe itself out! Some studies have suggested that the Permian–Triassic extinction was caused by microbes. For millions of years beforehand, environmental stress had caused conditions to deteriorate, with a combination of global warming and a slowdown in ocean circulation making it ever more difficult to replenish the oxygen which marine life drew from the water. According to this theory, microbes saturated the oceans with extremely toxic hydrogen sulfide that would have readily killed most organisms. Single-celled microbes survived, and indeed may well have prospered in that environment, but everything else was devastated. It took 10 million years for complex life to recover. And intelligent life, even if it avoids extinction by nuclear war, could develop a technology that causes its demise. Nanotechnology, for example. This is at the forefront of research and development in so many technical areas. How could it pose a risk? The term refers to engineering on the nanoscale level, which is billionths of a meter. This offers marvelous developments, but also introduces threats. On the drawing board, so to speak, is a proposal for a nanobot as a fundamental building block that can replicate itself. If self-replication were to get out of control, the population of nanobots would grow at an exponential rate. Let us say that a nanobot needs carbon for its molecules. This makes anything that contains carbon a potential source of raw material. Could nanobots extinguish life on Earth? Also, whilst there are many possible advantages to medicine at the nanoscale level, anything that enters the body also represents a potential toxin that could be difficult to eradicate. To be safe, we might have to control nanotechnology much as we control the likes of anthrax, namely by isolating it from the world at large. This is a problem that will face civilization in the next decade.

It is therefore thought unlikely that intelligence could develop on any planet that is subjected to extinction events on a frequent basis.

3 Planets and our Sun

The formation of planets is closely related to the formation of stars, but the physics is intrinsically different. Planets form in the final phase of a star's creation, using the material that is left over. As the Sun formed from an interstellar cloud some 5 billion years ago, the remaining dust and ice began to coalesce into billions of objects, each only several kilometers in size. These objects would have progressively collided, to create ever larger objects by the process of accretion. Mercury, Venus, Earth and Mars are rocky bodies that formed in the warm environment close to the Sun, and Jupiter, Saturn, Uranus and Neptune are gas giants formed in the cooler realm farther out. It seems that the gas giants formed before the rocky planets, and the asteroids are the small fragments that were left over. It is believed that whereas the gas giants were formed in several million years, the rocky planets required tens of millions of years. The gas giants are primarily hydrogen and helium, with only a trace of what astronomers call metals. The time available for the formation of gas giants was so limited because the radiation pressure from the new star would have blown the remaining gas away. In a gas giant, the heavier elements are located at the center.

Jupiter is 318 times the mass of Earth. In fact, Jupiter is more massive than all of the other planets combined. Although the rocky planets are insignificant members of the solar system in terms of mass, they are necessary for the development of life, so understanding the likelihood of there being Earth-like planets is key to assessing the probability of there being life in other stellar systems. A related issue, which we will discuss later, is how a Jupiter-like planet decreases the chance of an object such as a comet from striking a rocky planet and extinguishing complex life on its surface. To understand how Earth formed in the presence of a planet as massive as Jupiter, we need to know how Jupiter itself was formed. However, whilst the process by which a rocky planet forms seems to be agreed, the manner in which a gas giant forms is still disputed. The two leading theories can be characterised as bottom-up and top-down.

The bottom-up theory of bodies colliding and forming larger and larger bodies is the core accretion approach. To produce a gas giant, this theory requires that a rocky core with a mass at least 10 times that of Earth be formed very rapidly, to allow time for its gravity to begin to capture an envelope of hydrogen and helium. As the acquisition of this envelope is potentially a runaway process, the final mass of the planet depends upon how much time elapses before the newly formed star blows away the remaining nebula. If the process continues for long enough, the result is a sibling star rather than a planet. This bottom-up theory is favored by most planetary scientists, but because it proceeds very slowly it could imply Jupiter-like gas giants are rare in planetary systems.

The top-down approach is the disk instability theory, and draws its name from the fact that the angular momentum of a rotating interstellar cloud which is collapsing will reshape the initially spherical cloud as a thick disk. It postulates that clumps of gas and dust will form in the disk, and then interact owing to the intersection of random density waves in the disk. Where such waves reinforce each other, they form clumps with enough self-gravity to overcome the forces trying to rip them apart. The dust grains contained in such a clump settle to the center to create a rocky core which, once sufficiently massive, starts to grow a gaseous envelope. As this process runs so rapidly, it implies that Jupiter-like planets will be common in planetary systems.

The significance of this for SETI is that by protecting the planets in the inner region of a planetary system from comets, the existence of Jupiter-like planets increases the chances of a rocky planet hosting complex life. Understanding planet formation is therefore essential to the search for intelligent life.

Rocky planets

Rocky planets form by a process of accretion in the disk-shaped nebula of dust and gas that swirls around a forming star. It begins with individual dust grains colliding and sticking together. The growing conglomerates produce planetesimals ranging up to 1 kilometer in size. After millions of years, the colliding planetesimals assemble planets. The gravitational fields of the planets stir up the remaining planetesimals, causing them to collide violently and fragment in a process which ultimately reduces the material to dust. Although the planets sweep up some of the dust, this process is halted when the remaining nebula is blown away by the newly formed star, so there would appear to be an upper limit to the size to which a rocky planet can grow. It is also evident that rocky bodies would not be possible if it were not for the presence of dust in the nebula from which a star forms. More specifically, for rocky bodies to form, the nebula must contain elements heavier than helium. These heavy elements, which astronomers refer to as metals, were not created in the Big Bang. They are the result of fusion processes inside stars. The first stars were formed from nebulas that had no metals. The most massive stars burn most fiercely and expire soonest, and as they explode in supernovas they seed the interstellar medium with metals. Hence the metallicity of the universe increases with time. Successive generations of star have higher metallicity. Recent observations indicate a strong correlation between a star possessing a high metallicity and having giant planets traveling in close orbits. The presence of giant planets so close to these stars argues against there being Earth-like planets. Recall that in the solar system the rocky planets are close to the Sun and the giants orbit beyond. Perhaps rocky bodies are possible only for a narrow range of metallicity in a star. If the metallicity is too low then there will be no raw materials to form rocky bodies, but if it is too high

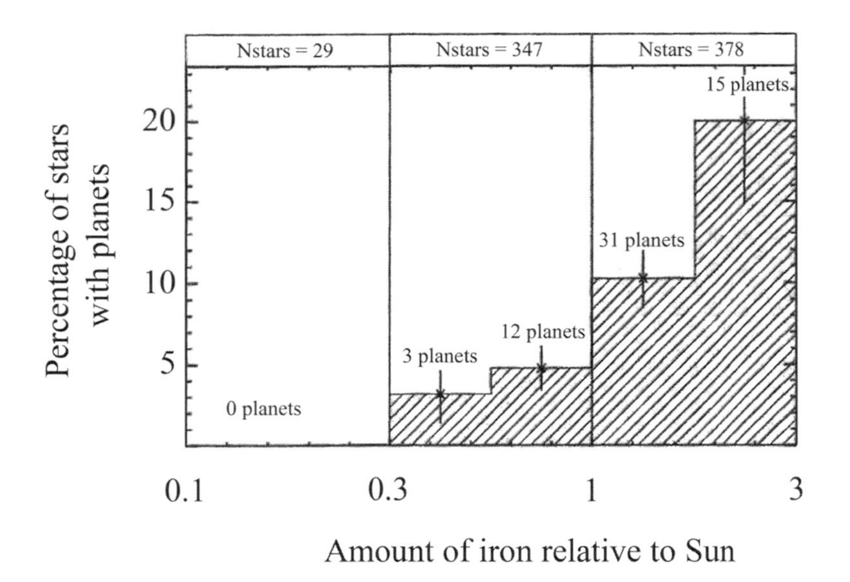

Figure 3.1. The occurrence of planets correlates with stellar metallicity.

it will lead to an abundance of giant planets. A study of extra-solar planets shows high-metallicity stars to have many more planets than low-metallicity stars. In fact, as many as 20 per cent of the high-metallicity stars studied have planets. Figure 3.1 shows the distribution of 724 planets in terms of metallicity in their host stars, as reported by Debra A. Fischer and Jeff A. Valenti. Furthermore, because the universe is 13.7 billion years old and the Sun is only one-third as old, it could be that three-quarters of terrestrial planets are on average 2 to 3 billion years older than Earth. This suggests that we are unlikely be the first advanced civilization in the universe. But constraints suggest that Earth-like planets are rare, which in turn limits the prospects for intelligence.

The case for intelligence being rare has been presented by Peter Ward and Donald E. Brownlee in their book *Rare Earth: Why Complex Life Is Uncommon in the Universe*, published in 2000. If they are right, there are serious implications for how and where we ought to search for alien signals. The authors cite many factors which contributed to the development of intelligent life on Earth, including the presence of a large moon and a very large gas giant orbiting farther from the Sun.

The Moon plays three pivotal roles in aiding intelligent life on Earth: (1) it causes lunar tides, (2) it stabilizes the Earth's spin axis, and (3) by a process of transferring angular momentum it progressively slows Earth's rotation rate. The most important factor seems to be the Moon's effect on Earth's axial tilt, which is measured relative to the perpendicular to the plane of the Earth's orbit around the Sun, known as the ecliptic. Obliquity is important because it varies the distribution of solar energy across the globe and gives rise to annual seasons. If the Moon were not present, the Earth's spin axis would wander in response to the gravitational pulls of the Sun and Jupiter, which would result in an unstable atmosphere and a chaotic climate. The monthly travel of the Moon around Earth acts to hold the spin axis stable. In fact, the angle of obliquity has strayed by no more than 1 degree from its nominal 23 degrees. If the Moon were much smaller or much further away then it would not have this effect. It simply happens to be just the right size. In terms of the solar system's satellites, the Moon has the largest mass relative to its host. Furthermore, if Earth orbited closer to the Sun or if Jupiter were either larger or orbited closer to the Sun, then the Moon's gravitational influence would be too weak to hold Earth's spin axis stable. Research suggests that a moderate obliquity is beneficial to life. Since the majority of a star's energy that reaches a planet heats the center of the hemisphere that faces the star, disagreeable climates occur if the axial tilt is either much shallower or much greater than ours. Although Mars has an obliquity of 25 degrees, comparable to ours, its two satellites are too small to stabilize it, and its obliquity would appear to have ranged between zero and 60 degrees in as little as 100,000 years, causing its climate to vary significantly and imposing severe stress on any life that may have developed there in the distant past.

Ocean tides are another important benefit to Earth from having a large Moon. The gravitational pulls of the Moon and the Sun each raise bulges in the ocean pointing towards and away from these bodies. When the Moon lines up with the Sun every fortnight the tides are at their maximum. Another effect of this interaction is to cause the radius of the Moon's orbit to increase and the rate of Earth's rotation to slow. Studies have suggested that when the Moon formed, it was only 24,000 kilometers away and Earth rotated in 5 hours. Today, the Moon is in a slightly elliptical orbit with an average radius of 380,000 kilometers and Earth rotates in 24 hours. Currently, the Moon is receding at about 4 centimeters per year. If it continues to do so, then in 2 billion years it will have insufficient influence to stabilize the obliquity of Earth's axis.

So, by this line of reasoning we may well owe our existence to the fact that Earth has an exceptionally large Moon.

Rare Earth summary

Whilst a number of conditions must be satisfied for a planet to have liquid water on its surface, it seems difficult to have the right amount of water for intelligent life to develop. For example, it would seem that although life is likely to originate in water it must then migrate onto land to develop technology, and an excess of water might result in a global ocean. The conditions required to produce just the right amount of water may imply that Earth is a very rare occurrence indeed. The first requirement is that the planet be at the right distance from its star for the temperature to allow water to be present on its surface in the liquid state. Next, this situation must remain stable for a very long time. Where does the water come from? Some will be acquired directly from the nebula as the planet accretes, and will be released by volcanic activity and condense when the surface cools. Not only would an impacting comet be an immediate threat to the survival of any life indigenous to the planet, the additional water it would deliver might threaten life's long-term prospects. We appear to have been lucky, in that not only is Earth at the right distance from the Sun to have liquid water on its surface, it has a large Moon to stabilize its spin axis and Jupiter has effectively swept away most of the comets left over from the solar nebula and thus reduced the frequency of impacts striking Earth sufficiently to preclude an excess of water and allow time for life to evolve intelligence.

Earth appears to be a low-probability event. It is what a statistician would refer to as a three-sigma (or greater) event. On a normal probability distribution, one sigma is one standard deviation from the peak value. Three sigmas would be three standard distributions and hence lie out on the tail of the distribution. In fact, Earth may even be a six-sigma event, far out on the extreme of the tail of the distribution with a four-in-ten-million chance. If this is the case, then within 1,000 light-years there would be few, if any, Earth-like planets.

Planets such as Earth are more likely to be in what is called the 'galactic habitable zone', identified as an annulus ranging from 22,000 to 30,000 light-years from the galactic center in which the stars are between 4 and 8 billion years old. Such stars would tend to have a metallicity suitable for the formation of rocky planets, and be far enough away from extinction events such as nearby supernovas to allow enough time for complex life to evolve. As most stars are not in this region, the majority of the enormous number of stars in the galaxy can be eliminated as candidates. When the reasonable requirement that for communication to be viable civilizations should be separated by no more than 1,000 light-years, the number of planets of interest to SETI drops much further. When we add the requirement that two civilizations must overlap in time, one can see how difficult it may be to find someone to talk to. We will explore all of these considerations in the next few chapters.

4 The many constraints on life

Whilst life may exist even in the harshest of circumstances on other planets, possibly even in our solar system, what is it about our situation that allowed complex life and ultimately intelligence to evolve. First, the Sun is a single star rather than a multiple-star system, it is rich in metals but not too rich, and it is in a relatively sparse part of the galaxy rather that in the center where there is a greater threat of extinction from a supernova or a gamma-ray burster. As for Earth, it orbits the Sun in a circular path with a stable temperature that has allowed liquid water to exist for billions of years. Although a planet can have a stable orbit in a binary star system if the components are sufficiently separated for the planet to rotate about only one star, or if they are so close together for the planet to orbit the pair of them, even if that planet were to start off in a habitable zone, any complex life that developed may well be extinguished by the changing conditions if the stars were of different masses and evolved at different rates. The galaxy contains globular clusters which can host up to 100,000 stars in a volume only 100 light-years across. In contrast to the disk of our galaxy, in which the stars travel together in essentially circular orbits, the extremely high spatial density of the stars in a globular cluster and the fact that they travel in random orbits presents a severe risk of gravitational interactions that would disturb the orbits of planets and deny the long-term stability required for intelligent life to evolve. In contrast, there are only 23 stars within 13 light-years of the Sun, making interactions unlikely. In the following pages we will discuss some of the constraints which imply that only a small number of stars within 1,000 light-years of us have a planet similar to Earth. The manner in which these constraints either reduce the probability of life developing or restrict its ability to evolve intelligence are also discussed.

It seems that in order for a planet not to suffer extremes in surface temperature, it must have an orbital eccentricity of less than 0.2, a rotational period of less than 96 hours and an axial obliquity of less than 55 degrees. To be habitable, a planet must have liquid water on its surface and a temperate climate. A thin atmosphere, or one that provides only moderate insulation, may be biocompatible but not habitable. An atmosphere that provides excessive insulation would create a greenhouse effect that would boil the surface water and end up making the planet as inhospitable as Venus. Finally, the planet must be volcanically and tectonically active in order to regulate the greenhouse gases, in particular carbon dioxide. The greater the mass of a planet, the longer it will remain geologically active. Earth releases its heat smoothly by the process of plate tectonics which continually renews the surface and precludes a build up of internal heat just beneath the surface. Although similar in size to Earth, Venus seems not to have plate tectonics, and fairly recently suffered a spasm of volcanism on a global scale. Mars, being much smaller, is inert, with an atmosphere of carbon dioxide that is too thin to generate a greenhouse and its surface is freezing and arid. Furthermore, for a planet to be habitable, its star must be more than about 0.75 solar masses. This is because small stars are less luminous, and the fainter a star the closer a habitable planet must be in order to gain the required energy. However, for a star as small as a red dwarf the planet will have to be so close that tidal braking will slow and eventually halt its rotation. Many factors that influence life are related to the rate at which a planet rotates, including diurnal temperature variations, the force of winds and the global climate. A slow rotation rate might strain the capacity of plants to survive a prolonged night. A more subtle reason has to do with the masses of the planets in such systems. Smaller stars will tend to possess less massive planets, and these will become inert sooner. Averaged across the time a star spends on the main sequence, low-mass stars will have fewer habitable planets than solar-type stars.

It is awe inspiring to look at the sky and realize that there may well be intelligent beings somewhere out there. The chance that they will be at about the same level of technology as us is very small – they are more likely to be either far behind us or far ahead of us. As Raymond Kurzweil pointed out in his book *The Singularity is Near: When Humans Transcend Biology*, published in 2005, our technological capability is developing at an ever-increasing rate. It is astonishing how primitive the technology a century ago seems, and impossible to fully imagine what we will have in 100 years from now. Our level of technology is bound to have an effect on how we search for extraterrestrial signals – for example, we started out using radio frequencies because that was a mature technology, whereas now we are considering lasers. What will we deem appropriate in the future? If 1 million years is divided into 100-year segments, this yields 10,000 segments. If we divide mankind's history into blocks of technological development each spanning 100 years, then it is at best 1:10,000 that an alien society is at the same level as us. The fact that we developed the ability to attempt to listen out for extraterrestrials in the last 100 years strongly suggests that anyone we find will be far ahead of us; if they are behind us, they will not be able to communicate across interstellar distances. Even applying criteria to limit the search, it will be like looking for the proverbial needle in the haystack, but the search is worth it because this will be a needle from which impossible dreams can be sewn.

Prokaryotes and eukaryotes

There is a major difference between simple life and complex life. Prokaryotes are single-celled organisms which comprise the molecules necessary for life surrounded by a membrane and cell wall. They are known to have been present on Earth as early as 3.5 billion years ago. Eukaryotes are much more complex multi-celled organisms, and developed 1.5 billion years ago. Living cells are classified into six 'kingdoms' depending on structure. Prokaryotes contain the kingdoms of Monera and Archaea. Eukaryotes have the kingdoms of Protista, Fungi, Plantae and Animalia, with the latter including the animal kingdom of which humans are a member. It was a huge evolutionary step from prokaryotes to eukaryotes, and a long series of steps from the simplest of eukaryotes to high forms such as animals. It is possible that prokaryotic life is common throughout the universe, and that eukaryotic life is rare. Complex life requires a benign environment to persist for a very long time. Macroscopic life such as animals are much less robust than microscopic life, and hence are more at risk of mass extinction. Therefore, life may evolve to a certain level over billions of years only for a major impact to wipe out all but the simplest life, thereby resetting the evolutionary clock. Animals require oxygen to sustain their high metabolic rates. It apparently required 2 billion years for simpler forms of life to infuse the atmosphere with sufficient oxygen to facilitate the development of animals. Is an oxygen-rich atmosphere a prerequisite for intelligent life?

Extremophiles

If the energy received by a planet from its star is not constant, then extremes of heat and cold could inhibit the development of complex life. Even if a planet spends most of its time in the star's habitable zone, periods of excessive heat might cause it to be sterilized. Nevertheless, in recent years research has shown life to be able to adapt to a much wider range of conditions than was originally thought. It exists in seemingly implausible environments such as deep underground in cold basalt, in gold mines, in hot springs on land, in hydrothermal vents on the ocean floor, and in sea ice. It had been believed that sunlight was the ultimate basis of life, but many of these forms of life have no need for light. They gain their energy from the dissociation of chemicals such as methane and hydrogen sulfide. The temperature increases as one descends into the Earth, and they draw their nutrients from the rocks they live in. Oxidized forms of iron, sulfur, and manganese can be nutrients, as can any organics trapped as the rock was deposited.

Many of these 'extremophile' microbes are bacteria, but the majority are a variety known as archaea. Not only do archaea live off chemicals which are toxic to other branches of the tree of life, most archaea cannot live in the presence of oxygen. As Earth originally had no free oxygen, this makes them candidates for being the 'root' of the 'tree of life'. It is possible that similar life forms developed independently on other planets and moons. However, although the presence of such microbes may be prevalent elsewhere in our solar system and in other star systems, it is difficult to imagine how this form of life could evolve into a civilization capable of transmitting signals across space. Nevertheless, prokaryotes gained the process of photosynthesis long before eukaryotes evolved, and it was their production of oxygen that changed the atmosphere into one suitable for animal life. Microbes degrade and remineralize organic material, produce methane, reduce sulfate to sulfide, and integrate molecular nitrogen. They drive the global cycles of carbon, nitrogen, phosphorus, and sulfur. To put it starkly, without microbes, complex life would not be present on Earth.

Stars and planets

Another factor that enters into the possibility of life developing to a complex level is the impact rate of comets and asteroids. The amount of debris left over in forming a planetary system determines whether the impact rate will be significant. The more bodies there are whose orbits cross that of a planet, the more likely it is to suffer an impact. A major impact may cause a mass extinction that comes close to requiring evolution to restart at a simpler level. Although it has suffered major impacts, Earth would appear to have a relatively low impact rate. It may be protected somewhat by Jupiter, which, by virtue of its size and position, serves to deflect objects entering our solar system. But how many Jupiter-like planets are in star systems with Earth-like planets? This may be another necessary condition for the development of advanced life. We can extend the requirement to include the star being in a neighborhood of the galaxy where the star density is sparse, encounters are infrequent and supernovas are unlikely. Not only do the stars in the galactic disk travel together in essentially circular orbits, they occupy essentially the same plane and hence do not repeatedly pass through the disk at steep angles that would cause gravitational interactions that would destabilize planetary systems. And, of course, we can eliminate variable stars, multiple-star systems, young hot blue stars, red giants, red supergiants, white dwarfs and exotic objects such as neutron stars from consideration as possible abodes of life. It should be noted, however, that the lowest mass extra-solar planets yet found are orbiting a neutron star.

In the near future, we will be able to infer from the spectrum of a planet orbiting another star whether it has an atmosphere and how this is structured, plus the range of temperature variation due to the axial tilt, rotation rate, etc. It therefore stands to reason that an advanced civilization will also be able to do this and limit their SETI effort to the most likely star systems: those with Earth-like planets which satisfy the necessary criteria. And since, as noted, any civilization attempting this is likely to be more advanced than us, they will be able to do their targeting on a much firmer basis than we are able today. Furthermore, the argument can be made that if there are so many constraints, there cannot be many planets bearing intelligent life. If this is so, then there is likely to be at most 100 planets within 1,000 light-years to send messages to, and probably less given that only 1 million of the 8 million stars within that volume are of the most promising F and G spectral classes. Hence it could be that there are only one or two other civilizations besides us, and they are either less advanced than us and cannot yet search, or are more advanced and have been looking fruitlessly for years.

Earth has another unusual property in terms of its metal content. It is accepted by space scientists that in addition to water, metals are critical to advanced life. Without heavy elements, and in particular iron, planets would not have internal heat sources or magnetic fields. Magnetic fields protect life from the rain of particles from space. Since a planet must have a certain amount of mass to develop a core of liquid iron and induce a dynamo effect, smaller planets might not have magnetic fields. Certainly Mars has no global field. And although Venus is as large as Earth, its rotation is very slow and it has no field. Even if a planet in the habitable zone close to a red dwarf is large, it is unlikely to have a magnetic field since its rotation will be tidally locked. But there is the interesting case of Mercury, which is a small planet whose rotation is very slow and yet does have a field, although this might be a consequence of the planet having a very large iron core and being close to the Sun. Finally, as they are present in organic matter, metals could also be a requirement for the development of animal life.

Water and plate tectonics

It is clear that there are many possible constraints on the evolution of complex life, but it is difficult to prioritize them. However, everyone agrees that water is crucial. As can be seen from comparing desert life with rain forest life, small changes in the availability of water in different regions of Earth have profound effects. It has been observed that a planet's water supply must satisfy four conditions: its water must be released from the interior onto the surface; it must exist for the most part as a liquid; it must be sufficient to sustain a large ocean; and it must not escape to space. It is an interesting fact that plate tectonics is involved in satisfying all four of these criteria. The temperature of the Earth's surface is dependent on several factors. Of primary importance is the amount of energy received from the Sun. However, the calculation must take into account how much of this energy is absorbed and how much is lost to space. In addition, the type and amount of greenhouse gases in the atmosphere will affect the temperature. A volcanically active planet will release a variety of gases into its atmosphere, but without a means of locking it back into the rock the carbon dioxide will build up. A large amount of carbon dioxide in the atmosphere will drive a runaway greenhouse, which is what made Venus too hot to support life. Without a mild greenhouse, Earth would be inhospitable to advanced life. Plate tectonics plays a crucial role in this. According to calculations by the Harvard–Smithsonian Center for Astrophysics, the greater a planet's mass, the greater the likelihood that it will undergo plate tectonics. But the process would appear to be confined to planets of a size ranging between one and two Earth masses. And a planet exceeding 10 times Earth's mass would probably have formed the core of a gas planet. The process by which rigid plates of a planet's crust are driven across the surface by convection in the mantle and crash into each other produces complex chemistry and, by melting rocks which contain carbon dioxide, releases this into the atmosphere. Meanwhile, the hydrological cycle absorbs carbon dioxide from the atmosphere and rainfall erosion of the mountains that are built up when plates collide produces sediments that return carbon dioxide to rock. This ongoing cycle on the geological timescale acts to stabilize the Earth's habitability.

A good discussion and reference regarding the issues of habitable planets can be found in Victoria Meadows work as chair at the University of Washington in charge of NASA's efforts on exobiology.

The recently developed technique of stellar seismology allows us for the first time to observe processes that occur deep inside a star. In 2006 the French space agency launched the COROT mission. Its telescope examined solar-like stars to measure fluctuations in brightness caused by acoustical waves generated deep in the interior which sent ripples across its surface. The pattern of these ripples enabled the precise mass, age and chemical composition of the star to be calculated. There may come a point where an advanced civilization can use a technique such as this to study the inner workings of planets of other stars to determine whether a planet operates plate tectonics in order to evaluate its prospect as a SETI target.

Oxygen

By volume the Earth's atmosphere is 78 per cent nitrogen, 21 per cent oxygen, some argon and 0.038 per cent carbon dioxide. Oxygen participates in most of the chemical reactions which release the energy required by complex life. Oxygen reacts with carbon compounds to produce water, carbon dioxide and energy. The energy utilized by living cells is stored in molecules of adenosine triphosphate (ATP). We require a constant supply of oxygen in order to maintain a reserve of ATP. Because oxygen is very reactive, its percentage in the atmosphere is close to the upper limit that is consistent with life today. The recycling systems of plants and animals have stabilized it at an equilibrium level – whereas animals take in oxygen and release carbon dioxide, plants do the opposite.

The fact that there was very little oxygen in the atmosphere early in the Earth's history posed no problem for the primitive prokaryotes. Once the photosynthesizing cyanobacteria evolved, they began to release oxygen. For a long time, this was simply absorbed by iron to produce the banded-iron formations of rock. Only when the iron was all oxidized could the fraction of oxygen in the atmosphere begin to rise. This transformation of the atmosphere was lethal to most of the prokaryotes, but some organisms evolved to exploit the higher metabolic rates that oxygen facilitated and they thrived. There is evidence that there was insufficient oxygen in the atmosphere to promote the development of animal life until about 2 billion years ago. Some 500 million years ago, at the start of the Cambrian Era, another major change increased the fraction of oxygen to a level more like that of today.

Mario Livio, author of numerous articles and books on astronomy, has produced a model of how a planetary atmosphere can develop to the point that it is capable of supporting life. There are two key phases. The first involves solar energy in the 100–200-nanometer wavelength range, at the short end of the ultraviolet part of the electromagnetic spectrum, dissociating water molecules to produce oxygen. Over an interval of 2.4 billion years this could have built up the fraction of oxygen in the atmosphere to 0.1 per cent of its present value. The second phase involves an increase in oxygen and ozone levels to about 10 per cent of the present value. This was sufficient to absorb ultraviolet in the 200–300-nanometer range and thereby protect two key ingredients of cellular life: nucleic acid and proteins. Life on dry land would be difficult, if not impossible, without the protection of an ozone layer.

It is clear that the fractions of oxygen and carbon dioxide in the atmosphere have changed substantially over the years. There appears to have been much higher levels of oxygen 300–400 million years ago, possibly as high as 35 per cent. At that level, forest fires would break out readily. It must have been a perilous time for animal life. On the other hand, the early land animals may have required such high levels of oxygen for their metabolisms to function. There have also been periods with reduced oxygen which would have favored organisms with less demanding metabolisms. It is clear, therefore, that instability in the chemistry of the atmosphere could inhibit the evolution of life capable becoming intelligent. In addition, if the crust of the planet is too thick or the mantle is too cool to convect, then plate tectonics cannot operate, and volcanism will cause a build up of greenhouse gases. The viable temperature range is much broader for prokaryotes than for higher forms of eukaryote. Without the process of plate tectonics, this narrow range of perhaps 5°C to 40°C could either be unattainable or unsustainable. The internal structure of a planet could therefore be fundamental to the evolution of intelligence.

In summary

The constraints on life on Earth are:

- right distance from the Sun
- right planetary mass
- plate tectonics
- Jupiter-like planet further out in solar system
- liquid water on surface
- stable planetary orbit
- large moon
- right axial tilt
- infrequent giant impacts
- right amount of carbon
- right atmospheric properties
- magnetic field.

As our investigations of astronomy and planetary science improve, we will learn more about these constraints. Some may come to be of major concern, others less so, and others that we do not yet recognize may become significant.

5 Why would anyone transmit to us?

There are three possibilities for why our SETI program to-date has failed to detect a signal:

- There are no advanced civilizations out there.
- We have not looked properly, or for long enough.
- They don't wish to send.

This book is about the second possibility. However, we should consider the other possibilities even although we are powerless to overcome them. If after many years of seeking a signal we remain unsuccessful, then we might reasonably conclude that we are alone. The third possibility begs the question of why an advanced civilization would choose not to attempt to seek out other life.

There are many reasons why *we* wish to receive a signal, including:

- We might receive practical information that helps our society flourish.
- We might gain new insights about major issues like wormholes and black holes.
- We might gain a deeper appreciation of ourselves as an intelligent form of life.
- It might lead to our involvement with more than one civilization, as a sort of 'Galactic Club'.

On the other hand, there are risks and potential negative effects, including:

- We might suffer from culture shock and lose confidence in our own future.
- Contact with powerful societies poses the risk of their dominating us.
- We might be wiped out.

However, such risks would most likely arise only if we were to reply to a signal and reveal our existence. And that is the crux of the matter: if they transmit seeking a contact and obtain a response, they put themselves at risk. Why should they accept this risk? Perhaps there is a pact between the civilizations which are known to each other, with each knowing that if it were to attack another it would be wiped out in response. This club may be reluctant to alert an up-and-coming civilization to its existence until the threat has been properly assessed. In fact, if we were to receive a signal from 200 light-years away we would probably think long and hard before we sent a reply. If we are lucky, there will be civilizations transmitting because they are so advanced that they have no fear of beings as technologically immature as us. But how much more advanced might they be? Our species originated only a few million years ago, and we began to develop societies several thousand years ago. It could be that a technological civilization lasts only a few million years. It seems unlikely that a civilization could last for a billion years. Our best hope would be to discover a civilization that was only slightly more advanced than us. We might not be able to relate to a truly ancient civilization; indeed, we might not even recognize it for what it is.

Another possible reason for a civilization to transmit is that it is concerned about an event predicted to occur in several thousand years, and is seeking a means of circumventing this disaster. Alternatively, they may wish to alert us to an impending catastrophe so that we could prevent it. It has been suggested that there is a 'galactic directive' adhered to by members of the Galactic Club that a civilization is contacted only after it reaches a certain degree of sophistication. Certainly an energy pulse in a highly directed beam that covers only our star system could be detected only if we had a large antenna and a sensitive receiver. Such a directive might be designed to protect technologically immature civilizations. If so, then we have just attained the status of being considered worthy.

Why would anyone bother to try to communicate with us when the time to elicit a response is so long? One possibility is that they are a knowledge-based society that faces long-term and complex issues, perhaps of a scientific nature, and believes that by sending us information we will reciprocate with something that will be of use to them.

But ultimately, civilizations might accept the risks of announcing their presence and transmit probing signals simply because they, as do we, yearn to know if there is anyone else out there.

Our SETI began with a bias for seeking signals at microwave frequencies due to the presence of the 'water hole'. This is the band from 1,420 to 1,640 megahertz that lies between the hydrogen spectral line and the strongest hydroxyl spectral line. The name was coined by Bernard Oliver in 1971 because the lines are the dissociation products of water, which is considered crucial to life. It attracted SETI because there is little natural noise in this part of the spectrum. Searches at water-hole frequencies have been pursued at the expense of other options, such as searching in the optical spectrum. In fact, there are excellent reasons for using optical wavelengths. First, of course, to broadcast at radio frequencies requires much more power than a laser and the antenna for an optical system is 1,000 times smaller than one for the microwave range. If the transmitting society were concerned about revealing its location, then it could either aim a laser directly at a star known to host a friendly civilization, or if it were conducting a search it could place a remotely controlled relay many light-years away from its home so that any response could be evaluated to determine whether to make direct contact. An unwillingness to broadcast 'Here We Are' may well explain our lack of success in finding signals at radio and microwave frequencies. In fact, it may be that we are the only ones dumb enough to reveal our presence by radiating radio waves indiscriminately.

Contact

Why might there be signals? Shouldn't an advanced society by this time have sent spacecraft or drones to investigate interesting stars? The answer is that the distances are so vast that to be even remotely realistic it would require a vehicle to be able to travel at a speed approaching that of light. As of now, we have no realistic means of achieving even a substantial fraction of the speed of light. A voyage to a star system 1,000 light-years away, at the boundary of the envelope considered for SETI, would take tens of thousands of years. Interstellar travel therefore appears unrealistic. Both the ability to achieve an appropriate speed and the time spent in flight are significant obstacles. Although the time dilation effects of Albert Einstein's theory of relativity would allow aliens traveling at near the speed of light to experience a flight time of a few hundred years, by the time they returned to their home planet tens of thousands of years may well have elapsed and their society would have changed dramatically. A further obstacle to traveling at near-light speed is that a means must be devised to protect the spacecraft from cosmic rays and strikes by interstellar dust particles. The difficulty of interstellar travel makes it more likely that a civilization's first contact with aliens will be by seeking and detecting electromagnetic signals.

This rationale is based on the proposition that nothing can exceed the speed of light. However, physicists have speculated on a type of subatomic particle called the tachyon which travels faster (indeed can only travel faster) than light. Tachyons are permitted by relativity, but there is no evidence that they exist. And even if they do exist, it can be argued that they could not be used for the purpose of communication. Discussion of the details of tachyon theory is far beyond this text, since it requires a deep understanding of quantum mechanics.

Michael Hart in 1975 reasoned that we are the first advanced civilization to arise in our galaxy. He offered four rationales for why there are no intelligent beings from outer space on Earth today: (1) sociological explanations – they have no wish to visit us, (2) physical explanations – they consider interstellar travel to be impractical and so do not try, (3) temporal explanations – they set off anyway and are still in transit, and (4) explanations arguing that they were once on Earth but are not now. Hart said that all reasons for there being no alien visitors must fall into one of these cases, and then offered strong arguments to suggest that there are in fact no aliens and that we are the first. In this book we will not try to validate or argue Hart's thesis, we merely point out that only a civilization that was convinced that it was the first would see no merit in seeking signals. On the other hand, it is often assumed that any civilization that develops the capability to explore the unknown will do so.

But here on Earth there are examples of societies that developed the capability to explore and did *not* do so. In the fourteenth century, China was the leading maritime country with a fleet of over 300 ships. Chinese admirals voyaged to the Persian Gulf and East Africa as China expanded its empire, then the emperor and leading admiral died and China not only ceased to explore but also forbade foreign trade and turned its attentions inward for centuries.

Another reason for there having been no contact comes from theories like parallel universes. If parallel universes exist, and there were a method for crossing between them, this would offer a far more interesting pursuit to an advanced civilization than simply trying to communicate with the likes of us. Far-fetched? Probably. However, it is hard to believe that even such a civilization would lack the curiosity to seek out societies in its own universe.

Panspermia

Whilst this book is not oriented towards the question of the origin of life, we must give this some attention in determining the likelihood of advanced beings sending us signals. There are a number of theories for how life on Earth originated, and one old idea that has recently regained support is panspermia. This proposes that life arrived from space. In one form of the theory, the building blocks for life arrived on bodies ranging in size from dust grains up to comets or asteroids, and the development of life occurred on Earth. In the other version microbes were transported through space and simply colonized Earth. The material could either have originated in our solar system or have arrived from interstellar space. It is possible that such microbes were transported not by chance but by a 'seeding' probe on a mission. In that case, there may be genetic commonality between civilizations in different star systems. If life on Earth arrived by space probe, there is no need for anyone to contact us - someone is watching us grow. Ultimately we may reach the level that allows us to meet those who seeded life here.

The Fermi Paradox and basic issues

The Fermi Paradox is the disturbing proposition that given the age of the galaxy, one would think it should be saturated with advanced civilizations by now. Yet we see no evidence of extraterrestrial intelligence. Any hypotheses on aliens must therefore seek to accommodate this observation. Whilst various theories to explain this have been offered, it is straightforward to plug numbers into the Drake Equation (which is covered in detail in Chapter 6) to indicate that there can be only a few civilizations co-existing in any time. In that case, the chances of their achieving communication are low - even if only taking into account factors such as knowing where to point an antenna and which frequency to listen on. By this argument, it is not surprising that we have heard nothing. But our search has not been rigorous. If after 100 years of serious searching we still have a negative result, we might subscribe to the Fermi Paradox.

There have been many SETI papers written that present reasonable arguments for different conclusions about the likelihood of civilizations transmitting. It is worth discussing some of the reasons for expecting a low probability of success in making a search.

First, it has been suggested that civilizations do not last long enough to overlap in time. The argument is that a civilization will inevitably collapse for one reason or another – perhaps after as little as 10,000 years, which is insignificant in terms of the age of the galaxy. On the other hand, if we rule out all forms of demise other than extinction events then a civilization could last a billion years, which is a substantial fraction of the age of the universe. Depending on the assumptions, the chances of civilizations co-existing are either negligibly small or almost a certainty. Other explanations for our failure to detect a signal are that we have not looked well enough or long enough or in the right way, and that other civilizations are afraid to transmit lest they reveal their existence. And, of course, there is the awful prospect that there is simply no one out there.

Let us consider further the proposition that civilizations are unlikely to last long enough to co-exist and communicate. This argument is an extrapolation of the fact that human civilization has suffered its ups and downs, and sometimes regressed for a while. It says that all civilizations must be unstable, and will destroy themselves by warfare or succumb to a natural extinction event. It is easy to play a numbers game showing that of a total of X stars, only Y could support complex life and of these Z would develop intelligent life, with only a small fraction overlapping in time. These would be so far away from each other as to make contact unlikely. For example, if one out of every million stars had intelligent life at some time, and if we divide their lifetimes into 10,000-year segments over 1 billion years, then only one other star out of 100 billion would host intelligence at that same time. The distance between two stars with intelligent life being present simultaneously can be expected to be so enormous that there would be little chance of their finding each other.

As a starting point to attribute a low chance of success for our SETI efforts, it was asserted that there was a low probability of other stars having planets. It transpires that planets are more common than thought, and this illustrates two points. First, not only is it likely that the assumption that rocky planets are rare is incorrect, but also, despite the weight of 'expert' opinion, any assumption without hard evidence stands a good chance of being wrong. (In fact, this has always been true of science.) It is irrational to base a decision on speculation based on an inadequate impression of the universe rather than upon data – even if hard data is sparse. One cannot justify *not* searching for signals by a succession of unproven assumptions designed to show the low probability of another intelligent civilization co-existing nearby.

The argument that civilizations may be reluctant to reveal their existence makes several unstated assumptions. One is that it is reasonable to fear your neighbors, but in the absence of faster-than-light ships there is no physical threat from a civilization light-years away. Second, it presumes that there is a rational reason for the recipient of a signal to attack and that the sender would not have the ability to deal with such a threat. Physics makes it highly unlikely that a ship could be accelerated to a speed approaching that of light. But if this is the concern, then it might be more rational to aim signals at stars that met the sender's criteria as potential abodes of intelligence whose technology was likely to be less advanced than that of the sender. That would be more understandable. Selective aiming implies utilizing a laser rather than a radio transmitter, because a laser beam is narrower.

In contrast, we are radiating radio and TV signals into space in a spherical pattern, available for anyone to pick up. We have been at it for the better part of a century, so in principle our presence is detectable from anywhere in a radius of 100 light-years. Some might argue that this puts us at risk. However, as discussed earlier, the power of a signal decreases with the square of the distance traveled, and it must be pulled out of the background of noise. A light-year is 9.46×10^{12} kilometers, so for ours to be detected would require either a suitably equipped ship to be nearby or a massive antenna at another star. The angle over which an antenna receives is proportional to its diameter, so it is theoretically possible to make a microwave dish 1,000 meters in diameter capable of picking up a random signal from Earth, but the likelihood of our radio-frequency and microwave emissions being detected is not great. Nevertheless, it is rather distressing to think we might be judged by an episode of *I Love Lucy*. The powerful microwave signals produced by air defense radars would be much easier to detect, but would carry little information.

The "we are alone" argument is based on assumptions of how life originated and evolved, and presumes unlikely events about which we have no hard evidence. The only way to falsify the statement is to find a signal. So a prerequisite to success is to search. However, the fact that absence of evidence is not evidence of absence means that the statement cannot be proved true. The entire exercise is a judgement call that reflects a society's willingness to pursue a potentially pointless exercise in the hope of proving otherwise. There are several hundred billion stars in our galaxy. A case has been made that other civilizations are likely to be older than us. This is for two reasons. First, a large number of stars in the galaxy are older than the Sun, some of them billions of years older. It follows that civilizations hosted by such stars will have arisen sooner. Second, it is plausible that some civilizations are long-lived. Our civilization is about 10,000 years old, and technologically competent for only a few hundred. If civilizations range in lifespan from several thousand to a million years, then if half are older than us and half are younger, even this weak argument suggests that some will overlap with us and hence there is a fair chance of our search for a signal being successful.

Longevity and number of civilizations

What can we really say about the lifespans of civilizations? Astronomer Ray Norris of the Australian Telescope National Facility has calculated as follows. The galaxies were made soon after the Big Bang, which was 13.7 billion years ago. The Sun was formed 5 billion years ago, and is half way through its time on the main sequence. Some of the first solar-type stars to form will therefore have evolved into red giants. Assume, therefore, that for solar-type stars it takes 5 billion years for an advanced civilization to arise. This allows a civilization 5 billion years before its star evolves. By this simple reasoning, there would not have been any civilizations during the first 5 billion years of our galaxy's existence. Thereafter, assuming a constant rate of star formation, the number of civilizations would have risen linearly. Their median age is therefore the median of the civilizations that started between 5 and 10 billion years ago, which is 1.7 billion years. Ignoring other factors, any civilization that we detect would probably be well over a billion years more advanced than us. In reality, this figure is likely to be much lower due to the many ways in which a civilization might be wiped out. Furthermore, Norris estimates that there is less than one chance in a thousand of the advanced civilization being less than a million years older, hence any civilization we detect will be far more advanced than us.

If Norris is anywhere near being correct, it becomes even more evident how unrealistic it is to search only at radio frequencies. Any part of the electromagnetic spectrum would be available to such a civilization, and the choice would depend entirely upon factors about which we can only guess. Perhaps they would wish to limit contact to civilizations at least as advanced as themselves. Maybe this would involve utilizing a method that relies on a principle of physics about which we know nothing. If this involved faster-than-light transmission it might enable bidirectional interstellar communications to match our present intercontinental communications networks.

One of the major issues in deciding how many civilizations are out there is their longevity, which is limited by events that can destroy intelligence on a planet. In the case of Earth, mass extinctions occur about every 200 million years, with the range of possible causes including a comet or asteroid strike, a supernova explosion within 50 light-years, and a gamma-ray burster in our part of the galaxy. Yet life has been evolving here for almost 4 billion years, absorbing setbacks to the higher forms of life and resuming evolution. In statistical terms, the sooner that intelligence develops after an extinction, the longer it has to flourish before the next threat. On the other hand, if civilizations are unlikely to live as long as a million years, as presumed in the argument above, then this weakens the conclusion that any society is likely to be far more advanced than us and helps to explain the silence.

6 The Drake Equation and habitable planets

In 1960 Frank Drake devised an equation which expressed the number of intelligent communicating civilizations in the galaxy as the product of a number of probability factors, viz:

$$N = N^* f_p n_e f_l f_i f_c f_L$$

where:

N^* is the number of stars in the galaxy
f_p is the fraction of stars that host planets
n_e is the number of planets per star capable of sustaining life
f_l is the fraction of planets in n_e on which life evolves
f_i is the fraction of f_l that evolves intelligence
f_c is the fraction of f_i that communicate
f_L is the fraction of the planet's life during which the civilization actively communicates.

Whilst this equation identifies the important factors, it offers no clue to the value of each contributing factor. In some cases we can make educated guesses, but others are unknown and hence subject to wide variances depending upon whether one feels optimistic or pessimistic on the issue. We will discuss each factor and identify what we really know – which, truth be told, isn't much!

The number of stars in our galaxy (N^*) has been estimated at between 100 billion and 400 billion. We are fairly confident of this, so the variance is not as large as for some of the other factors.

The fraction of stars that host planets (f_p) is estimated to be between 20 and 50 per cent. The accelerating rate at which we are discovering extra-solar planets indicates a high probability factor.

The number of planets per star that are capable of sustaining life (n_e) is estimated to be between 0.5 and 2. As most of the planets discovered so far are gas giants even larger than Jupiter, and are orbiting very close to their parent stars, the evidence to achieve a better number is not yet available. However, observational methods better suited to spotting small rocky planets are being developed and so we should soon be in a position to improve our estimate of this factor.

The fraction of planets in n_e on which life evolves (f_l) is really unknown. We gain little insight by using the solar system as a guide. In fact, plausible arguments can be made for a value close to 100 per cent and for almost zero. One calculation is that life-bearing planets can exist only around stars inside an annulus around the galactic center that is quite narrow. This argument is based on the fact that as stars in the disk of the galaxy orbit the center in more or less circular orbits they pass through the spiral arms, which are self-perpetuating density waves laced with clouds of gas and dust. The proposition is that life can only exist for the length of time that the parent star spends crossing the gaps between spiral arms. The spiral is much tighter towards the center, so, it is argued, life can exist only further out. The number of stars in the specified annulus is only 80 million. Given the value of N*, this would indicate f_l to be much less than 1 per cent; to be specific, it would be 1 in 5 million, which is an amazingly low percentage. If this is correct, then it narrows dramatically the portion of the galaxy that one should search.

Next, we estimate the fraction of f_l that actually evolve intelligence (f_i). If we are generous in choosing f_l, should we be so here also? Again, guesses can be made that range from a very low percentage to almost 100 per cent. We simply do not know. If the solar system is any guide and only Earth proves to have life, then the value of f_i is 100 per cent, but this will diminish if there proves to be any life on Mars or in the ocean beneath the icy crust of the Jovian moon Europa.

Estimates of the fraction of intelligences that try to communicate (f_c) range from 20 per cent to 80 per cent. However, one line of argument is that if civilizations are reluctant to reveal their existence then the value might be very low.

Lastly, we have the lifespan of the civilization (f_L), which must overlap in time with other civilizations for communication to occur. Choosing a value for this factor has a substantial influence on the outcome, but we have very little data to go on. Can a civilization remain stable for tens of thousands of years? Hundreds of thousands of years? Millions of years? Billions of years? Can we expect our civilization to last for many thousands of years, or will we either self-destruct or be wiped out by a natural disaster? The equation expresses f_L in terms of the fraction of time during which the planet is habitable. The Sun formed 5 billion years ago. In another 5 billion years it will evolve off the main sequence and become a red giant, at which time Earth will become uninhabitable. This permits us to set an upper limit to f_L. As this also applies to other solar-type stars, the fact that many such stars are older than the Sun suggests that even the longest-lasting civilizations have come and gone.

If a society signaling to us is more advanced than we are, this can only be because it started earlier and lasted a long time, or started about the same time and advanced more rapidly. It can reasonably be expected that the kind of natural catastrophes that befell Earth in the past would also afflict planets in other systems, and that a number of civilizations that signaled to potential neighbors died off before we were capable of detecting their signals. It is possible to make back-of-the-envelope calculations. A civilization could have started a billion years earlier than us because its star formed a billion years earlier than the Sun, and if the civilization survived it would inevitably be more advanced than us. If they advanced at the same rate as us, and technology increments of 100 years are used, there are 10,000 such segments in a billion years. It is open to question whether a civilization that started at the same time as us could have advanced at a significantly faster rate.

Vladimir Kompanichenko recently attempted to calculate the average lifetime of a civilization using the idea of the global cycle of human civilization. He considered a group of active systems that develop cyclically, starting at their origin and running through their peak of development to expiry. The following stages exist in the cycle of the total system:

- Growth, increase in size
- Internal development
- Maturity, stationary state
- Aging.

Kompanichenko analysed each stage, estimated the time for each, and calculated a value of 300,000 years. Whilst some of the assumptions may not be true, we do not have a better number.

We can readily see, therefore, that the value of N, the number of communicating civilizations, is highly dependent upon assumptions about which we can only guess. If we are optimistic we can calculate a high number, but if we are pessimistic we can come up with a low probability.

For example, if we take the fraction of stars that host planets as 20 per cent, the number of planets per star able to sustain life as 1, the fraction of planets on which life develops to be 10 per cent, the fraction that evolves intelligence to be 5 per cent, the fraction that attempts to communicate to be 20 per cent, and the time available to a civilization to communicate to be 10,000 years in relation to 10 billion years of the habitable lifetime, then we calculate the number of communicating civilizations in the galaxy at any given time to be 20. However, we can change this to only 2 simply by reducing the fraction of planets that develop life from 10 per cent to 5 per cent, and the fraction that evolve intelligence from 5 per cent to 1 per cent. Figures 6.1 and 6.2 illustrate a low number and a high number of communicating civilizations at any given time. Of course, it is relatively easy to choose values that make the number of communicating civilizations less than 1, to support the argument that we are alone. On the other hand, as in Figure 6.2, we can choose values that say that there should be thousands of civilizations, and the increasing evidence for extra-solar planets offers hope that there are indeed many civilizations communicating.

Webb's approach to the constraints on intelligence in the universe

Stephen Webb, in his 2002 book *If the Universe is Teeming with Aliens ... Where is Everybody?*, uses the Eratosthenes method to analyze the number of stars that might host advanced civilizations, and reaches the conclusion that we are probably alone. His book provides detailed explanations of each constraint in the

N^* = 100 Billion

f_p = 10%

n_e = 1

f_l = 10%

f_i = 1%

f_c = 20%

f_L = 10^{-6}

Then N is ~2

Figure 6.1. The Drake Equation solved for a low number of communicating civilizations at the same time.

N^* = 400 Billion

f_p = 50%

n_e = 2

f_l = 20%

f_i = 10%

f_c = 50%

f_L = 10^{-5}

Then N is 40,000

Figure 6.2. The Drake Equation solved for a high number of communicating civilizations at the same time.

eight steps of his analysis. Our purpose here is to point out that it becomes more and more likely there are not a large number of good candidates, and therefore that a targeted approach is necessary.

Webb's constraints are:

- For a star to possess a viable planetary system, it must reside in the galactic habitable zone.
- Only stars similar to the Sun should be considered as hosting *life as we know it*, because stars of spectral classes O and B are too luminous and expire too soon, and stars of classes K and M are too dim.

- Life as we know it requires a rocky planet that remains in the continuously stable habitable zone around its star for billions of years.
- For life as we know it to emerge, conditions on the planet must be suitable.
- There must be a low probability of a planet suffering a disaster that snuffs out life.
- There is the question of whether plate tectonics and the presence of a large moon are necessary for the evolution of complex life.
- On some planets life evolves intelligence.
- In order to participate in interstellar communication, a civilization must first develop the necessary technology.

As Webb admits, assigning probabilities to many of these constraints is no more than a matter of making educated guesses, with different people doing so differently. It should be noted that Webb's analysis does not consider the issue of a civilization choosing not to reveal itself.

It could be that as an advanced civilization accumulates more and more knowledge, it might perform such an analysis and conclude that there is unlikely to be anyone with whom to communicate, and hence opts not even to search for signals. Certainly, the incentive to continue searching must diminish as the distance in light-years between potential advanced civilizations increases.

Another way to consider the probability of there being an advanced civilization trying to contact us can be described in five steps, each of which requires a number of conditions to be met: (1) being in an advantageous place in the galaxy, (2) having the right kind of solar system with a planet similar to Earth that permits life, (3) the evolution of intelligence, (4) the conditions that allow transmitting across interstellar distances, and (5) the willingness to reveal themselves by signaling in an effort to elicit a response.

For us to have a chance of intercepting alien signals, all of the conditions required by each step must be satisfied. It is not so much the specifics of the conditions that is daunting, but their number. If we simply assume there to be at least 10 conditions to be met and argue there is no better than a 10 per cent probability for each, then since there are 8 million stars within 1,000 light-years and the cumulative probability is one in a billion of satisfying all the conditions, we must conclude there is unlikely to be anyone sending signals within that volume of space. Of course, this is simplistic because the probabilities will differ from 10 per cent, some being higher and others lower, but in reality there are significantly more than 10 conditions to be considered.

What kind of conditions will enhance the likelihood of intelligent life developing? We will discuss a number of conditions without assigning specific probabilities. As our understanding of the universe increases, hopefully our analysis of this issue will improve.

The first step in the analysis relates to residing in a place in the galaxy that is not thick with stars and interstellar gas and dust. In a region with a high star density, the possibility of adverse gravitational interactions and cataclysms such as supernovas will be greater. A related condition is that the region in which the stars that we are interested in formed in interstellar clouds that were rich in the metals necessary for life. As regards any particular planetary system, one of the crucial conditions is the distance of a planet from its star. The planet must be close enough for the surface temperature to be conducive to life, but not so close as to suffer tidal lock with one hemisphere being baked and the other frozen. It is debatable whether a tidally locked planet the size of Earth would retain an atmosphere in such conditions. Certainly the surface environment would pose a significant challenge to life.

We should bear in mind that we are looking for life that is sufficiently complex to have developed a technology capable of signaling across interstellar distances. This excludes forms of life that are unable to advance. An example of this is the bacteria recently found in a gold mine deep below the Earth's surface that lives in complete isolation, devoid of light and oxygen. It gains its energy from the radioactive decay of uranium in the surrounding rock, and has genes that cause carbon to be extracted from dissolved carbon dioxide and other genes to fix nitrogen from the surrounding rocks. Such microbial life may well be able to live on planets which are substantially different to Earth, and may, in fact, be commonplace, but it could never evolve sufficiently to become capable of communicating across space.

To provide the time required for intelligence to develop on a rocky planet close to a star, it may be essential that there be a Jupiter-like planet further out whose gravity deflects comets and asteroids, reducing the risk of an impact resetting the clock on the evolution of complex life.

An additional requirement is that the rocky planet has sufficient mass to retain an atmosphere. Those that do not have atmospheres are arid deserts. Furthermore, some atmospheres have a chemistry that would be toxic to life as we know it.

It has been recognized in recent years that the process of plate tectonics is critical to life, since it cycles greenhouse gases. In addition, the motions of the plates around the globe encourage biological diversity. This derives from the fact that the interior of the Earth is sufficiently hot for the silicate mantle to convect. Electrical currents in the liquid iron core generate a dynamo effect that provides Earth a magnetic field, and this shields the planet from the energetic charged-particle radiation of the solar wind. Although from time to time the field has reversed its polarity, with its strength falling to zero during the transition, this would appear to have had minimal effect on life. That said, the last magnetic reversal occurred long before the advent of human civilization and the presence of a distortion in the field known as the South Atlantic Anomaly could indicate that the next reversal is imminent. If so, it is sure to have a disruptive effect on us, if only because our compasses will cease to work for a while and will then point the wrong way.

It has also been suggested that a large moon like ours might be important for the development of advanced life, because it would stabilize the rotational axis of the planet and in turn moderate the climate.

There are a number of factors to be considered in selecting stars in a search for signals. In broad outline these are:

- the type of star
- the distance of a planet from the star
- the composition of the planet's atmosphere.

As previously noted, astronomers discuss stars by their spectral class, using the designations O-B-A-F-G-K-M. Class O are the hottest, with a surface temperature of 50,000K, and class M are the coolest at about 2,000K. The Sun is class G, and about 5,700K. It is a straightforward matter to calculate the surface temperature of a rocky planet at any distance from its parent star. However, this will be for a planet that has no atmosphere. The existence of an atmosphere will change the surface temperature. If Earth had no atmosphere, its surface would be –18°C. Greenhouse gases in the atmosphere have increased the surface temperature sufficiently for water to exist in a liquid state, which is a prerequisite for the development of life. Whilst Venus would naturally be warmer by orbiting closer to the Sun, it has a runaway greenhouse that increased the surface temperature to 475°C. A planet's proximity to the Sun is often conveniently stated in terms of an Astronomical Unit (AU) which corresponds to the average radius of the Earth's orbit. On this scale, Venus is at a heliocentric distance of 0.72 AU. Mars is too small to have a thick atmosphere, and at 1.52 AU is freezing. This situation is reminiscent of the story of Goldilocks and the three bears, with one bear's soup being too hot, another's being too cold and the third being just right.

The chemical process of life requires certain heavy elements to be present. Only some stars have what astronomers refer to as metals, by which they mean elements heavier than helium. It has been established that the metals were created by nuclear reactions inside stars, in particular in the massive stars which explode as supernovas. The metallicity of the interstellar medium therefore increases with time. The nebulas from which the earliest stars with masses similar to the Sun formed would not have had the metals to produce rocky planets. In fact, a survey (Figure 3.1) found that low-metallicity stars do not even possess gas giants. Life could develop only when planets were formed which had sulfur, phosphorous, oxygen, nitrogen and carbon. One theory is that only young stars such as the Sun itself could have planets capable of originating life. If this is so, then ours might be one of the earliest civilizations. If intelligent life is very rare, then it is likely we are the first.

One factor to bear in mind is that the receiving apparatus for an electromagnetic signal may have a limited range. That is, we might choose not to invest the effort to look further than about 1,000 light-years, in which case only a small fraction of the potential civilizations communicating in our galaxy would be within the range of our detector. This would significantly reduce the Drake Equation to a much lower value. For a galaxy that is 100,000 light-years across, a detector limit of 1,000 light-years would provide a sample of 1 to 5 million stars, which would substantially reduce the chance of our SETI effort being successful. The rationale for setting a range limit is that a sender's apparatus would need so much more power for the signal to reach us, and the travel time would be so great that bidirectional communication would be impracticable even between long-lived civilizations. If after a methodical search we cannot find anyone within 1,000 light-years, we might lose our enthusiasm to look further.

The number of candidate stars

Astronomer Margaret Turnbull devoted serious study to finding stars that look as if they might be habitable, and has recently reported that an initial screening turned up 17,000 candidates, of which a handful passed a more detailed screening. Several of her criteria related to the age of a star. She concluded that it must be at least 3 billion years old. Variable stars that are prone to energetic flares are generally too young to meet Turnbull's criteria for habitable planets. Also, stars exceeding 1.5 solar masses do not live long enough to create habitable zones. She also considered the issue of metallicity. If the spectrum of a star does not show a high level of iron in its outer envelope, the nebula from which it formed is unlikely to have had enough heavy metals to form planets. To satisfy Turnbull, a star had to have at least 50 per cent of the iron observed in the solar spectrum. And, of course, she only accepted stars that were on the main sequence of the Hertzsprung–Russell diagram.

At present, the Sun is our only guide to the type of star that can host an Earth-type planet on which intelligent life can evolve. If we examine the nearby stars, we find that the Sun has several special features. It is a solitary star, whereas most have one or more siblings. Stable planetary orbits are much easier to achieve in solitary star systems. If it is at the right distance, a planet in a circular orbit can remain within its star's habitable zone. A planet in a multiple-star system may well be in an elliptical orbit that crosses in and out of the habitable zone.

The Sun is among the most massive 10 per cent of stars in our neighborhood, so it is neither too cool and dim, nor so massive that it will burn itself out before life can develop and evolve to produce an oxygen atmosphere capable of sustaining animals and ultimately intelligence. Furthermore, the Sun appears to have about 50 per cent more metals than other stars of its age and spectral class, but only one-third of their variation in luminosity. This is fortunate, firstly because heavy elements are required to make rocky planets and also because major flare activity on a star could irradiate life on an orbiting planet with hard radiation. As no star is exactly like the Sun as far as we can tell, astronomers have been studying which factors are essential for life.

Turnbull identified only 30 possibilities for habitable zones within 100 light-years of the Sun. Her criteria included kinematics, spectral class or color, metallicity, rate of rotation, and its characteristics in four-color photometry and at X-ray frequencies. If there are so few candidates within 100 light-years, then any advanced civilization capable of signaling to us is likely to be further away.

Even allowing for the fact that the number of candidates will increase with the cube of distance, there would still be only about 16,000 candidates within a radius of 800 light-years. In Chapter 16 we will discuss methods of searching using a targeted approach capable of dealing with such a number and also solve the transmitter/receiver directionality problem.

As we learn more about extra-solar planets and the requirements for life, we will be able to tighten the constraints and reduce the number of candidate stars. When we consider all the factors required for there to be an advanced civilization transmitting signals, the odds are that anyone we might hear from will be at least several hundred light-years away. Of course, if they perceive the distances to be too great, they might strike us off their list! The fewer the number of candidate stars, the more likely it is that such beings will use directed beams rather than broadcasting.

It is possible that advanced civilizations originate only on planets around stars that have moderate-to-low mass, since they remain on the main sequence long enough to facilitate the requisite biological evolution. Stars of less than 0.5 solar masses can be rejected because a planet in the habitable zone will be so close to its star that it will become tidally locked before life can develop. Although stars of more than 1.5 solar masses may well have planets which originate life, such stars probably do not spend sufficient time on the main sequence for this to develop intelligence, so they can also be rejected.

Habitable zones are defined by the range of distances from a star at which water can exist as a liquid on a planet's surface, taking into account such additional factors as the planet's surface gravity and the density and chemistry of its atmosphere. The Sun's habitable zone currently extends from 0.95 AU to 1.37 AU, with the radius of the Earth's orbit being 1.00 on this scale (Figure 6.3). Main sequence stars brighten as they age, causing the habitable zone to migrate outward. It has been calculated that the Sun must have increased in luminosity by 30 per cent since it 'switched on'. The continuously habitable zone for a star represents the overlap of the habitable zone over two widely separated points in its history – i.e. over a period of several billion years. In the past 4.6 billion years the continuously habitable zone for the Sun was 0.95 AU to 1.15 AU. This constraint tightens the distance at which a planet can orbit its star and offer long-term support for life. The width of the current habitable zone is 0.42 AU, whereas the width of the continuously habitable zone is only 0.20 AU. This can be interpreted to mean that a planet with a heliocentric distance of between 1.15 and 1.37 AU could not have developed complex life because the planet has not yet spent sufficient time in the zone.

Special factors for Earth

Special factors that might apply include our position with respect to the galactic habitable zone. One of the Sun's characteristics is that its orbit around the center of the galaxy is significantly less elliptical than those of other stars of similar class

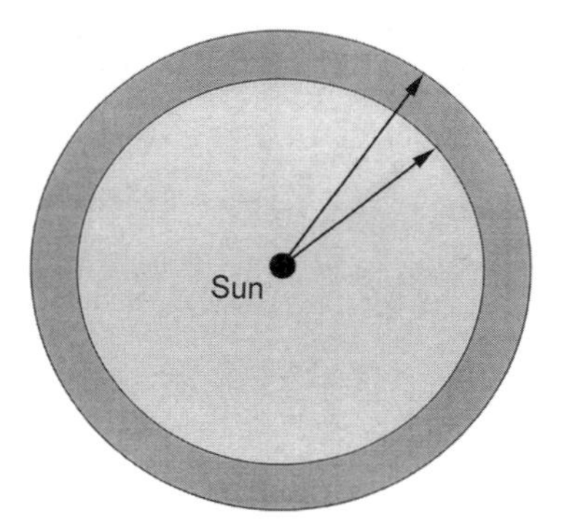

Figure 6.3. The habitable zone of the solar system currently ranges from 0.95 to 1.37 AU. When the Sun was weaker, early on, the zone extended out only to 1.15 AU.

and metallicity. Hence the Sun never enters the core, where the star density is higher and the threat to life from supernovas is greater. The small inclination of the Sun's orbit precludes abrupt crossings of the disk which could disturb the Oort Cloud and result in Earth being bombarded with large comets that might extinguish complex life. Our galaxy makes one rotation around its axis in about 250 million years. Interestingly, the Sun orbits very close to the 'co-rotation radius', where the angular speed of the stars matches that of the spiral arms. By reducing the frequency with which the Sun passes through an arm, this limits the Sun's exposure to massive young stars in the arms going supernova. It could be that life on Earth was able to evolve in complexity only while the Sun was in the relatively benign environment in between two spiral arms. It is estimated that fewer than 5 per cent of stars have this kind of galactic orbit.

The primary task of NASA's Kepler mission, launched in 2009, is to detect Earth-type planets around other stars. The term 'Earth-type' was defined to mean a rocky planet with a mass of 0.5 to 2.0 times that of Earth, corresponding to 0.8 to 1.3 Earth radii. The mission will probably also find planets up to 10 Earth masses, or 2.2 Earth radii, but our current understanding of how rocky planets form suggests that those with less than half the mass of Earth would not retain an atmosphere at all and those exceeding twice the mass of Earth might attract sufficient hydrogen and helium to gain an atmosphere that would not be conducive to the development of life. And, of course, rocks of more than 10 Earth masses would probably become the cores of gas giants. Hence the belief is that only rocky planets in the 0.5 to 2.0 Earth mass range are likely to give rise to advanced life.

The idea of a habitable zone was devised for roughly circular planetary orbits, but an eccentric orbit might spend some time in the habitable zone defined for a circular orbit. The eccentricity of an orbit states its departure from circular. An eccentricity of zero is a perfect circle. The Earth's orbit is very close to circular at 1 AU, but if it were to have an eccentricity of 0.3 our planet would vary between 0.7 and 1.3 AU, approaching the Sun as close as the orbit of Venus and receding as far as the orbit of Mars. Because the abundant water in the oceans has a very high capacity to absorb heat, the Earth would be slow to heat up during the brief perihelion passage where it would be traveling at its fastest, and this stored heat would serve it well while it was flying its slow aphelion. Modeling shows that during this annual cycle the surface temperature would neither become so hot nor so cold as to rule out life. It is difficult to draw a firm conclusion, but for eccentricities exceeding 0.3 the difficulties facing life would be formidable.

Choosing SETI candidates

In seeking signals from an alien civilization we cannot look everywhere, we have to make assumptions and choices. But we must fully understand the assumptions made. The most important assumption is that for intelligent life to develop, it needs a stable star system. The time that a star spends on the main sequence depends on its mass, in that more massive stars have hotter cores and therefore exhaust their nuclear fuel more rapidly. The Sun is expected to spend a total of 10 billion years on the main sequence.

In addition to a star residing on the main sequence, a number of assumptions are required to select the best SETI candidates. The distance of the star from the Sun has to be a major factor, since a civilization is likely to start by signaling to nearby stars rather than to those many hundreds of light-years away. But if it should turn out that intelligence is very rare, there may well be nobody signaling in our neighborhood. The number of candidates increase rapidly with distance, so if after applying various criteria we identified 200 candidates within a radius of 100 light-years, the fact that the volume of space increases with the cube of its radius would mean there would be 800 within 200 light-years.

The Earth is not near the center of the galaxy, but we can get an estimate of how many stars in total there are within a given radius of us. If there are 400 billion stars in the galaxy whose disk is 100,000 light-years across, the distance from a star near the center to one on the rim is 50,000 light-years. The distance from a star on the rim at one side of the disk to a star diametrically opposite is 100,000 light-years. But as noted earlier, there are reasons to believe that communication over such enormous distances is impractical. It is likely that any intelligence which was signaling would target the stars in its immediate vicinity prior to looking further afield. Although we also make the assumption that the star density is approximately equal throughout the galaxy, we know that this is not true because the stars are more closely packed in the center. Figure 6.4 shows that the Sun is located near the outskirts of the galactic disk. Observations show that there are 21 stars within 12 light-years of the Sun, and 50 stars within 16.3 light-years. The stars in the galactic disk in the Sun's vicinity are about 8 light-years apart on average. A statistical analysis suggests there are 1 million stars within 500 light-years. The Cyclops report estimated that there are at least a million stars on the main sequence within 1,000 light-years. However, it is clear that only a range of spectral types are appropriate. Clearly the Sun is suitable. But

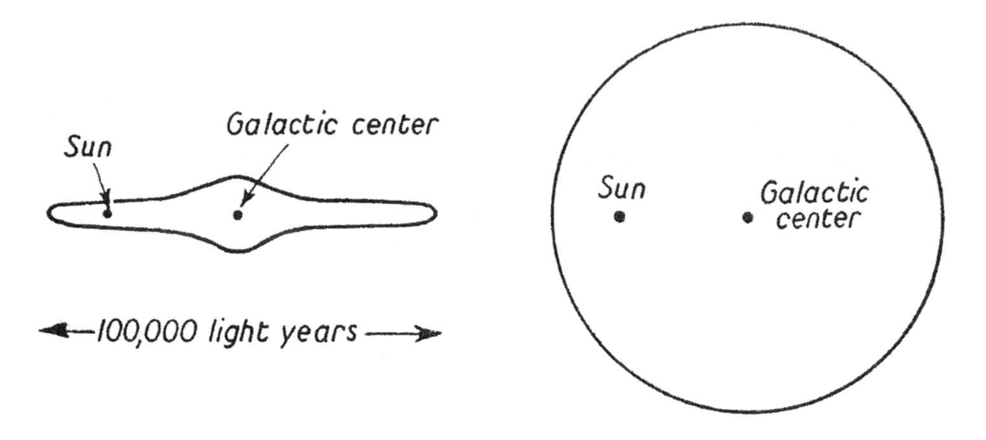

Figure 6.4. The galaxy has a spherical 'core' 30,000 light-years in diameter and a 'disk' 100,000 light-years across that is at most 10,000 light-years in thickness. The location of the Sun is shown in a side view (left) and a position above the galactic pole (right).

how narrow is the selection? Perhaps conditions on Earth are near the limit for intelligent life to develop? Other assumptions can also influence the selection process. For example, it is debatable whether binary stars should be included, since a planet in such a system may well suffer temperature extremes that would not be conducive to life.

Solving the Drake Equation can yield either a low number or a high number of communicating civilizations at any given time, but one might at first think that with a total of 40,000 communicating civilizations (as in Figure 6.2) we should detect a signal provided that we look in the right part of the electromagnetic spectrum.

However, it can be argued that even with such a high number it might be a difficult job. If the truth is a much lower number, the task is truly like looking for the proverbial needle in a haystack. With just a few civilizations trying to communicate in a galaxy that is 100,000 light-years in diameter, in all likelihood the civilizations will be separated not by several hundred light-years but tens of thousands of light-years. Even narrow-beamwidth lasers will require lots of transmission power in order to provide an adequate signal strength upon arrival. The less prevalent intelligence is within the galaxy, the further a signaling civilization may have to probe to attract our attention. If we assume that everything remains the same except the power of their transmitter, they would have to increase this by a factor of 10,000 to provide the same signal strength upon arrival after a distance of 100,000 light-years in comparison to 100 light-years. At a certain point, the inevitable dispersion of the beam would impose a further increase in the power. In any case, if a civilization is that far away, some of the arguments for why a society would communicate lose sense. Will anyone wait to expect an answer if it takes 10,000 years to receive a reply? Civilizations 100 light-years apart may have good reasons to attempt to exchange information in the hope of assisting their own society. The only way that civilizations 10,000 light-years apart could communicate meaningfully would be to employ some form of faster-than-light signal. In fact, it is likely that no civilization would bother to seek counterparts on the opposite side of the galaxy. If there are only a few civilizations capable of communicating, and these are distributed widely across the galaxy, then they are unlikely to find anyone to talk to within a radius of several hundred light-years, making true communication barely feasible.

If we now impose an arbitrary maximum distance upon a search, it is not the total number of civilizations in the galaxy that is pertinent, it is this number multiplied by the fraction of the stars in the galaxy which are to be searched. Hence, if we choose 1,000 light-years as the maximum then the sphere of interest is a very small fraction of the galaxy. Although the core of the galaxy is spherical, we are more interested in the disk hosting the spiral arms. The galaxy is about 30,000 light-years thick in the center, and 10,000 light-years thick further out. Given the position of the Sun in the disk, the volume enclosed by a sphere of 1,000 light-years radius is not much greater than one millionth of the galaxy's volume. As volume is a cubic function of radius, in Figure 6.5 we illustrate the situation using a cube for simplicity. The fraction of the volume of the galaxy that will be searched is the small cube inside the much larger cube. If we take the optimistic case of 40,000 civilizations signaling, is there cause for optimism that we will readily find one in our sample? Let us analyze this. If we consider the galaxy in terms of cubes with each side 1,000 light-years in length, then to a first approximation there are 300,000 such cubes in the galaxy. If we now divide

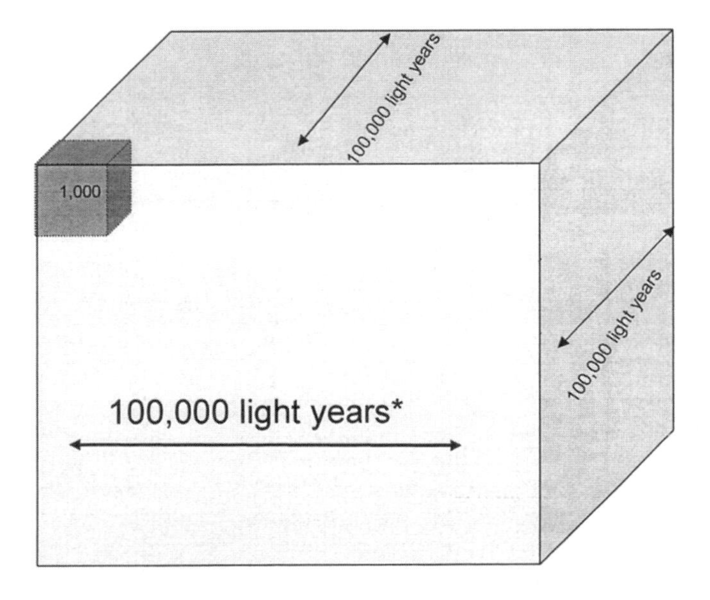

Figure 6.5. A 1,000-light-year cube represents only one millionth of the volume of the galaxy.

40,000 civilizations by 300,000 cubes, this gives 0.13 civilizations per cube. When we use Poisson statistics to calculate the likelihood of two civilizations being present in the same box (i.e. someone else and us) the result is much less than 1 in 100. This prompts the thought that neither those people who argue that we are alone, nor those who say the galaxy is teeming with intelligence, are entirely right. The aliens are out there, but the galaxy is so vast that they are unlikely to be within 1,000 light-years of each other. With 40,000 signaling civilizations in the galaxy, the probability of there being one within a cube that is only 100 light-years on a side is 1.3×10^{-4}, and the likelihood that we share such a cube with someone else is on the order of 10^{-8}. If the civilizations are presumed to be evenly distributed across the galaxy, then the simple act of applying a maximum radius to the search severely reduces the total number of communication civilizations produced by the Drake Equation. However, civilizations are unlikely to be evenly distributed throughout the galaxy.

In 1961 the German radio astronomer Sebastien Von Hoerner suggested that the most likely distance at which we are likely to find another civilization is 1,000 light-years. Although he had little data on which to base his analysis, he argued that we ought to presume that "anything that seems unique and peculiar to us is actually one of many and is probably average". Thus he assumed that the Sun is an average star and that Earth is an average planet. In the absence of evidence to the contrary, the assumption that we are average has the highest probability of being right. But on the other hand we do not have the slightest idea of how wrong we might be.

The discussion above is not intended to be definitive in any way, but it does show the difficulties of mounting a search. It is reasonable to conclude that intelligence is likely to be out there, but will be difficult to find.

Working in conjunction with its European counterpart, the French space agency launched the COROT mission in 2006. The 30-cm-diameter telescope will examine solar-like stars for a slight dimming as a planet crosses in front of its parent's disk. It is sensitive enough to detect rocky planets with diameters several times that of Earth, and may prove capable of finding Earth-sized planets. The Kepler mission was launched by NASA in 2009. Its 95-cm telescope will stare at a patch of sky to simultaneously monitor the brightness of 100,000 main sequence stars in search of transits. The science team expects to find about 50 planets of the same size as Earth, 185 of about 1.3 times the Earth's radius, and 640 of about 2.2 times Earth's radius. It further expects about 12 per cent of the stars to possess multiple planets, but most of these are expected to be gas giants.

Within a few years, therefore, we should be in a much better position to evaluate the structural diversity of planetary systems. This will improve our ability to choose a SETI target list, as well as provide a better idea of how many candidate stars there really are. Specifically, the data will:

- determine for a wide variety of stars the percentage of terrestrial and larger planets that are near the habitable zone
- determine the distribution of sizes and shapes of planetary orbits

- estimate how many planets there are in multiple-star systems
- determine the reflectivity and masses of the planets
- identify additional members of each discovered planetary system
- determine properties of those stars that have planets in the habitable zone.

Meanwhile, ground-based telescopes have been busily finding extra-solar planets. Several hundred have been found that are much larger than Earth, but these are not likely SETI candidates. However, as the methods improve, smaller planets are being discovered. In April 2007 the Geneva Observatory in Switzerland spotted a planet of a red dwarf called Gliese 581 that is about 20 light-years away from the Sun. The planet is about 1.5 Earth radii and about 5 Earth masses. It orbits 15 times closer to its star than Earth does the Sun, but red dwarfs are typically 50 times dimmer than the Sun so it is not roasted. Then in May 2008 a team led by David Bennett of the University of Notre Dame in the USA announced finding a planet of 3 Earth masses using the gravitational microlensing method. The star is a brown dwarf located some 3,000 light-years from the Sun, and the planet orbits at about the same distance that Venus does the Sun.

As new space telescopes and larger ground-based telescopes search the sky, we can expect planet discoveries at an increasing rate. A good way to keep up to date with discoveries is the http://planetquest.jpl.nasa.gov website.

Part 2: The basics of space communication

7 Where to look in the electromagnetic spectrum?

Based on our present knowledge of physics, if we are to detect a signal from space it will be as electromagnetic waves, somewhere in the spectrum between gamma rays at one end and wavelengths longer than AM radio transmissions at the other. This is because only gravitational and electromagnetic waves can travel through a vacuum. Sound waves, for example, need a material (air, water, etc) to vibrate. The light we see from the stars has traveled trillions of kilometers through the vacuum of space. As we also know, geometry makes the intensity of an electromagnetic wave fall off with the square of the distance traveled. Hence if the distance is doubled, the area reached is quadrupled and the signal strength is quartered (Figure 7.1). If a transmitter has a certain power, then it is advisable to make the beam as narrow as possible in order to maximise the signal at the receiver. It turns out that the higher the frequency of transmission, the easier it is to narrow the beam. This is why a laser appears as a sharp narrow beam of light, whereas a radio signal fans out over a wide angle.

Whilst electromagnetic energy can be thought of as waves, at short wavelengths these can also sometimes act like particles. This aspect of electromagnetism is called the 'wave-particle duality'. The frequency of a wave

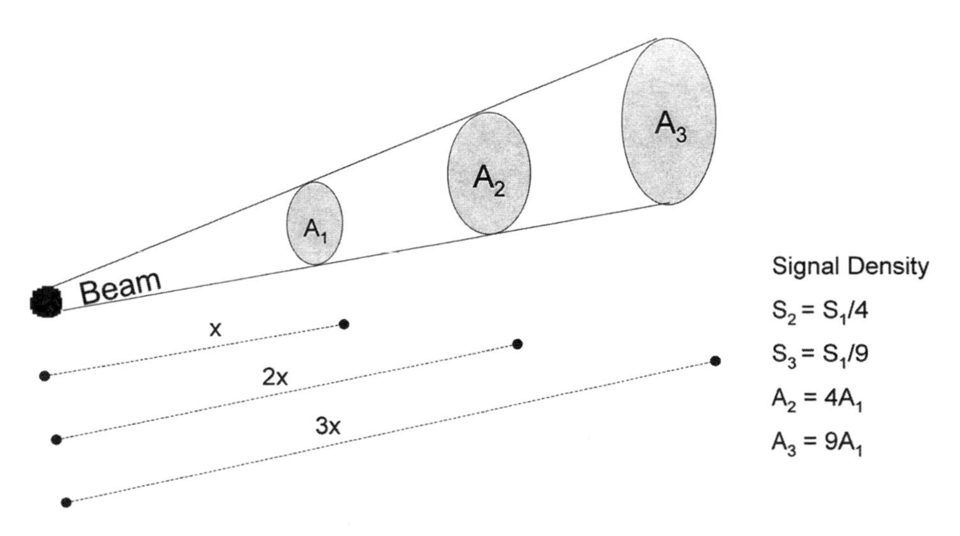

Figure 7.1. Signal density reduces with the square of distance because the area over which the energy is distributed increases with the square of distance.

is how fast the peaks of a wave travel past a given point, and the wavelength of a wave is the distance between those peaks. Clearly, at long wavelengths the frequency is low and at short wavelengths the frequency is high. For electromagnetic waves in a vacuum the product of the wavelength and the frequency is the speed of light, and the energy increases with the frequency. In most situations, high-energy photons such as X-rays behave more like particles than waves. In fact, in the quantum realm mass is a concentrated form of energy, so all matter exhibits both wave-like and particle-like properties. The idea of duality dates to work by Christiaan Huygens and Isaac Newton in the seventeenth century, with the former conducting experiments which perceived light as wave-like and the latter as particle-like. Who was right? By the end of the nineteenth century, diffraction and interference experiments were seen as definitive proof of the wave nature of light. But problems emerged a few years later. Albert Einstein's analysis of the photoelectric effect in 1905 could only be explained by light having particle-like properties. In the photoelectric effect, shining a light on a metal sometimes yields an electric current. As expected, the strength of the current was proportional to the intensity of the light, but it also proved to be dependent on the color of the light such that, depending on the metal used, the current would switch off for light longer than a certain wavelength. For example, no matter how bright a red light might be, there would be no current, whereas even a faint blue light would yield a current. This was inexplicable by wave theory. Einstein therefore inferred that the electrons that were escaping from the atoms of the metal were able to receive electromagnetic energy only in discrete amounts, and he postulated the existence of quanta of light energy in the form of particle-like photons. In the photoelectric effect, only photons above a threshold energy could liberate an electron. Blue photons (high frequency) could do so, but red photons (low frequency) could not. This explains why photomultipliers perform well in the visible, but as we move into the infrared the quantum efficiency drops abruptly to zero. The advent of quantum mechanics in the 1920s showed light to be neither a wave nor a particle, but something that can display the properties of either depending on the situation.

Gravitation also acts across space, but it is much weaker than electromagnetism. Electromagnetic signals are much easier to generate and to detect than gravitational waves, making them much more likely as a means of communication. We have yet to build a functioning gravitational wave detector, but if we had one it would likely detect only astronomical events.

Perhaps, the most surprising thing to lay people, is the fact that radio frequencies and microwaves are only a very small portion of the electromagnetic spectrum. This misconception has led to the public's belief that SETI is a comprehensive program which, after almost half a century, has had no success. Basic physics texts give the impression that radio frequencies and microwaves account for a substantial part of the electromagnetic spectrum. But one hertz of frequency is the same bandwidth whether it is at radio frequencies or in the ultraviolet, and hence to fully appreciate the spectrum it must be displayed on a linear frequency scale (Figure 7.2). Now we see that the SETI effort has been

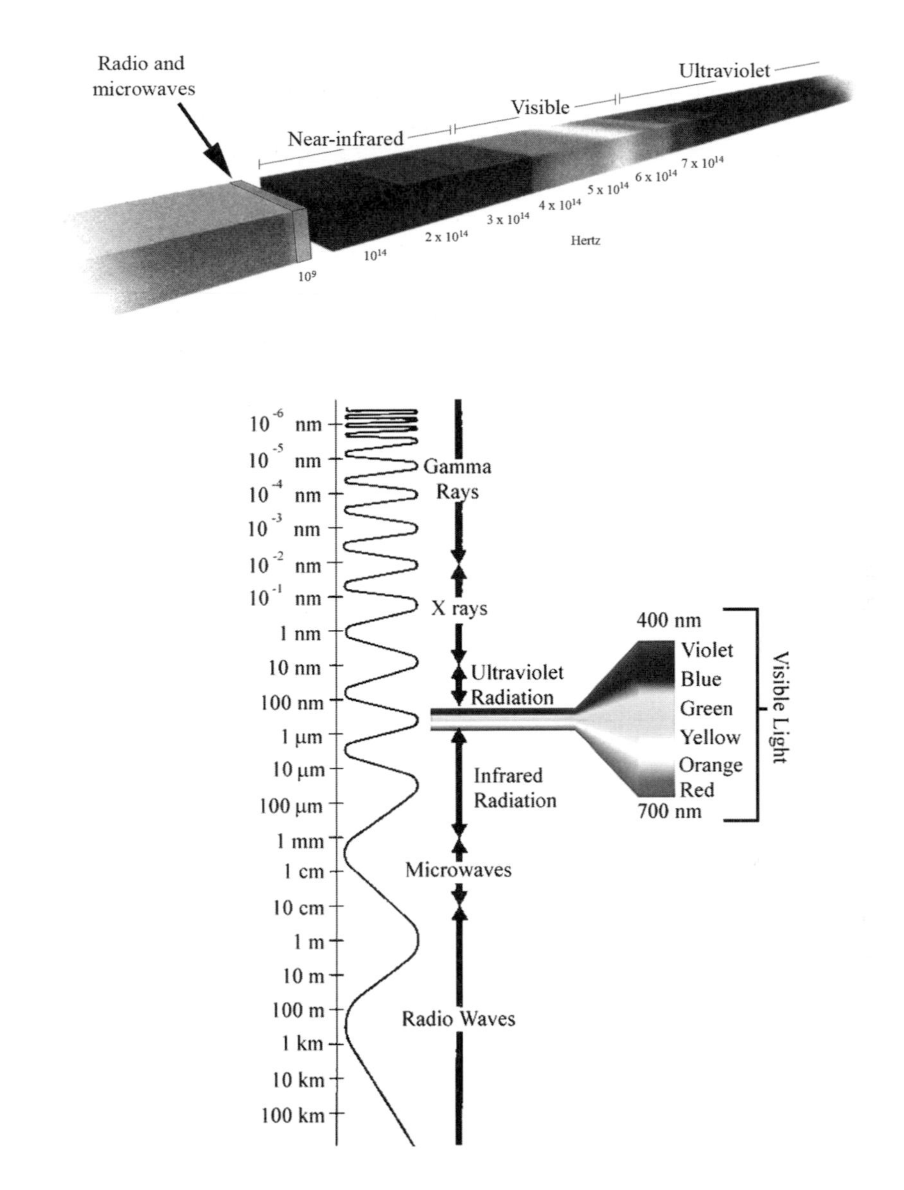

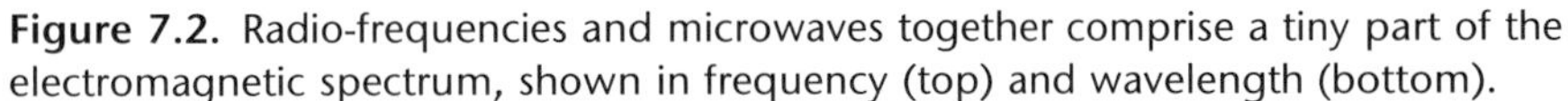

Figure 7.2. Radio-frequencies and microwaves together comprise a tiny part of the electromagnetic spectrum, shown in frequency (top) and wavelength (bottom).

confined to *less than one tenth of one thousandth* of the electromagnetic spectrum. As an analogy, imagine you are put on a desert island and told that a great treasure is buried there. Unfortunately, you have few tools and the easiest place to dig represents one tenth of one thousandth of the total area available. You convince yourself that it is the right place to dig because if it is the easiest place for you to dig then it would have been the easiest place for the person that buried

Wavelength	Frequency (Hz)
0.1 micrometers	3×10^{15}
1.0 micrometers	3×10^{14}
10 micrometers	3×10^{13}
100 micrometers	3×10^{12}
1 millimeter	3×10^{11}
10 millimeters	3×10^{10}
100 millimeters	3×10^{9}
1 meter	3×10^{8}
10 meters	3×10^{7}
100 meters	3×10^{6}
1 kilometer	3×10^{5}
10 kilometers	3×10^{4}
100 kilometers	3×10^{3}

Figure 7.3. The inverse relationship of wavelength and frequency for electromagnetic waves.

the treasure. But if that person had an earthmover, then the treasure could be anywhere. Anyone sending signals across interstellar space is likely to be more technologically advanced than us. For us to examine only radio frequencies and microwaves simply because we know how to do so, creates the illusion that SETI is being attempted on a comprehensive basis and has thus far been fruitless. The truth is, owing to the very limited part of the electromagnetic spectrum that has been explored, SETI has never been comprehensive. It is not that radio frequencies and microwaves ought not to be explored, rather that they constitute such a small part of the available frequencies. In later chapters, we will argue that the optical spectrum is a more sensible option. The problem is that the false impression created by efforts so far seems to be impeding a more realistic effort from gaining the necessary public support. Like the treasure hunter, many of the people examining radio frequencies know that other areas are deserving of exploration, but pursued the easier option first.

As noted, for electromagnetic waves in a vacuum the product of the wavelength and frequency equates to the speed of light of 3×10^8 meters per second (Figure 7.3). The energy of a photon is the frequency of its wave multiplied by Planck's constant, which is essentially a scale factor between the quantum and human realms and has a value of 6.6×10^{-34} joule-seconds. The

violet	380–450 nm
blue	450–495 nm
green	495–570 nm
yellow	570–590 nm
orange	590–620 nm
red	620–750 nm

Figure 7.4. Visible wavelengths.

total energy received by an instrument is the number of photons received, multiplied by the energy of a photon at the frequency to which the instrument is tuned. Obviously for a given received energy, many more photons will have been received at lower frequencies than would be so for higher frequencies. We will discuss this in much more detail when we consider how noise limits the detector sensitivity.

As a point of reference, an AM radio tuned to a frequency of 1 megahertz receives at a wavelength of 300 meters. The microwave range runs from 300 megahertz to 300 gigahertz. Beyond this, the Earth's atmosphere is effectively opaque until it becomes transparent again at infrared frequencies. The highest frequency microwaves are also known as millimeter waves because at 100 gigahertz the peak-to-peak wavelength is 3 millimeters.

The portion of the electromagnetic spectrum that our eyes use is referred to as the visible range, with wavelengths ranging from 380 to 750 nanometers (Figure 7.4). The human eye is far more sophisticated than all the apparatus constructed to operate at longer wavelengths. In fact, for over a century technology has been slowly working its way up the frequency range from power generation at 60 hertz to an AM radio transmission, to microwaves, to millimeter waves and more recently to lasers in the infrared and visible. It is generally harder to do things at higher frequencies, because it requires a deeper understanding of physics and because minor imperfections in the dimensions of the apparatus can cause serious problems. The fact that the eye is far more sophisticated than anything that we have built for ourselves has been noted by others. The emphasis that we wish to make here, and will do so throughout the book, is that surely it is reasonable that any advanced society attempting to communicate would use wavelengths they know that an advanced society such as ourselves would recognize to be more precise. In pondering this, consider why we see in the visible rather than at lower frequencies. First, the peak-wavelength of the Sun's output is in the visible spectrum, and the Earth's atmosphere is transparent at such wavelengths. Also, higher frequencies allow greater spatial resolution – we would have difficulty distinguishing a lion from a deer using millimeter-wave sensors. Clearly, evolution would favor sensors in this part of the spectrum. Optical telescopes were developed to aid the human eye by magnification. However, no further progress could be made until electromagnetism was discovered and the underlying principles understood. It is significant in this regard that Einstein received his Nobel Prize not for the theory of relativity but for his interpretation of the photoelectric effect. Our technology is still striving for shorter wavelengths. Consider the CD-ROM/DVD business, where we turned to blue laser light because the shorter wavelength permits more data to be written onto a DVD.

If we were attempting to gain the interest of the inhabitants of another star system, to use the transmitted energy as efficiently as possible we would confine the beam in order to restrict it to the zone in which complex life was most likely to exist. In the case of the solar system, an electromagnetic signal takes about 16 minutes to span the diameter of the Earth's orbit, and about 30 minutes to span the habitable zone (as defined in Chapter 6). If we are interested in solar-type stars, then once we know a star's distance we can calculate a beamwidth which will span its habitable zone and tolerate minor misalignments or beam jitter. This can be adjusted for hotter or cooler stars. The signal must arrive at the target star with sufficient energy to be detectable by anyone making a serious effort to detect such a signal – by which we mean they have a big antenna and a sensitive receiver. We will return to the meaning of 'big' in this context later.

Let us now reverse roles, so that it is another civilization that is signaling and the solar system is a potential target. They will probably know much more than we do at present about which stars make the best SETI candidates. Having compiled their list, they will signal each in turn. Even if they spend only a few days (as measured in our time) on each star, this task may take years. Every so often, the transmitter returns to look at each star. Because the round-trip will probably be hundreds of light-years, it would not be a big deal to transmit to the same star for several days every few years. It would be logical to send information that tells the recipient when and how to look for a follow-up signal. Hence, upon receiving a signal we might be told that we have until a given time to prepare. For example, the initial signal might tell us that a more substantial signal will be sent at a given time several years hence, for which we will have to build an antenna of a size that we would be unlikely to have built without a firm requirement. Meanwhile, the sender has a schedule for when to turn their own antenna to each star on their list to discover whether the probing signal has elicited a response.

Antenna gain and the value of directivity

As we do not know the direction from which a signal will come, it may seem that we ought to utilize an omnidirectional antenna which draws in signals equally from all directions, but this would actually make reception harder because the antenna's gain would be 1, which is the minimum possible. Not only has the signal strength been diminished by the interstellar distance traveled, the antenna offers no amplification. A directional antenna might have provided a gain of 100,000. For us to obtain that same received-signal strength using an

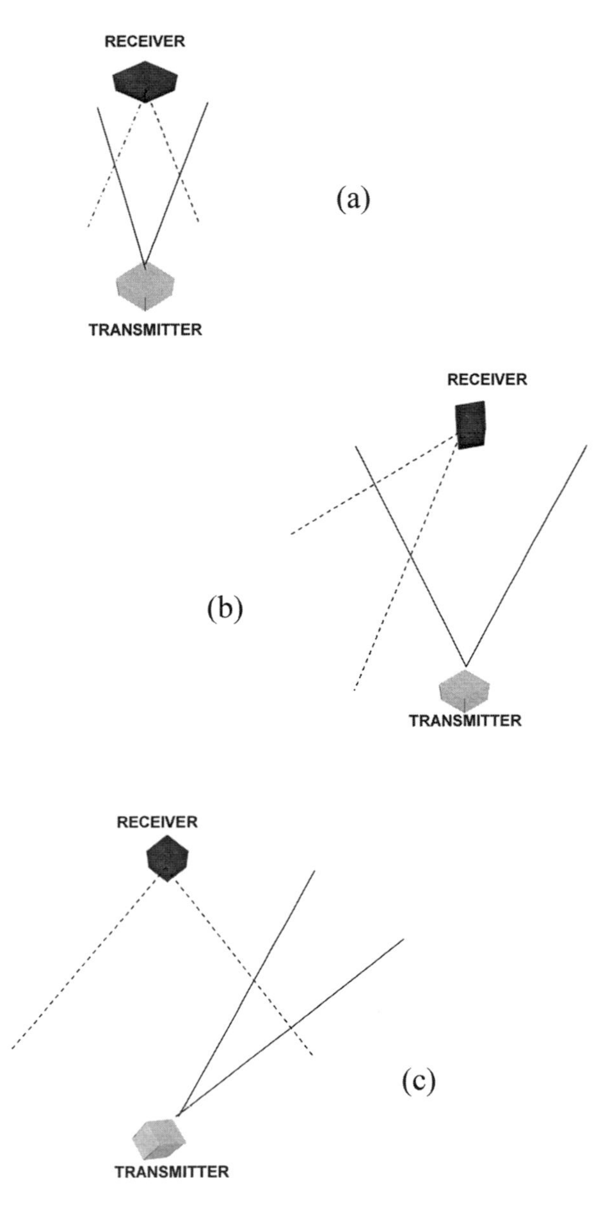

Figure 7.5. Transmitter and receiver directionality: (a) with the transmitter and receiver pointing at each other, (b) with the transmitter pointing at the receiver but the receiver not pointing at the transmitter, and (c) with the receiver pointing at the transmitter but the transmitter not pointing at the receiver.

omnidirectional antenna would require the transmitter to be 100,000 times more powerful, which would place an unreasonable expectation on the sender. The laws of physics therefore argue for using a high-gain antenna, but its directionality means it must scan a succession of small parts of the sky to examine each candidate in turn. Figure 7.5 shows the difficulty of attempting an uncoordinated communication using a pair of scanning systems (for simplicity we ignore the travel time of the signal). In Figure 7.5a we see a transmitter and a receiver pointing at each other as required. More likely situations are a receiver looking away when the transmitter is pointing at it, and a receiver pointing at a transmitter when the transmitter is looking away (Figures 7.5b and 7.5c respectively). Since there is no way to synchronize (unless it is a follow-up communication for which the timing has been specified), there is the highly important question of how to scan a receiver for a near-certainty of being pointed at a source at a time when that source is transmitting in our direction.

SETI efforts that do not take into account the requirement for both the transmitter and the receiver to scan are unlikely to succeed. As an illustration, consider a simple 3 by 3 cell pattern with a candidate star in each cell. The receiver devotes an equal amount of time to looking at each cell. Let us say that the transmitter is able to scan more rapidly and can cover 16 cells (which will be different than those the receiver is examining) in the same time that the receiver devotes to a single cell. If the Sun is on the transmitter's list and the receiver is simultaneously examining the cell which contains the transmitter, then (again ignoring the travel time for the signal) they will find each other. We can extrapolate this idea to the many thousands of candidates, and explore the parameters. In Chapter 13 we discuss in detail specific examples of viable systems which will ensure that the transmitter and receiver will point at each other simultaneously.

Obviously, the more sensitive the receiver, the easier it will be to detect a signal. Receiver sensitivity at radio frequencies has advanced as far as it can go, having run up against limits imposed by physics, and the only way to improve sensitivity is to build larger antennas. At microwaves, we face fundamental noise issues that cannot be eliminated. This noise is known by various names, including blackbody noise, Brownian motion, and thermal noise. Thermal noise is an apt name since the noise is directly related to the temperature of the receiver as measured on the Kelvin scale. One degree on the Kelvin scale is the same as on the Centigrade scale, but the zero is offset to –273°C. Absolute temperature is denoted by the letter T, and nothing can be colder than absolute zero. The other parameters required to define the noise are Boltzmann's constant, k, whose value is 1.4×10^{-23} joules per degree, and the range of frequencies accepted by the receiver, known as the bandwidth, B. The product of k, T and B is the noise that is caused by the random movements of atoms and molecules. As k is constant, the noise can only be cut by reducing T and/or B. For a bandwidth of 1 hertz and the most sensitive receiver chilled down to 3K, this noise is 4.2×10^{-23} watts. Although this is trivial compared to a 100-watt light bulb, which emits 100 joules of energy per second, the signal sought might be barely discernible above this noise. There is no way for the capability of these receiver systems to be significantly improved. The only solution is to increase the antenna's collection area in order to pull in more signal.

At optical wavelengths, where SETI is now underway, there is also good receiver sensitivity. Single photons can be detected. The energy of a photon is the product of h, Planck's constant, and f, the transmitted frequency (rather than the bandwidth). At one photon per second, the power at the near-infrared frequency of 3×10^{14} hertz would be a mere 2×10^{-19} watts, which is one-thousand-million-trillionth of a 100-watt bulb and an incredible sensitivity. We do not really get that in the visible range since not every photon is detected, but we can come fairly close. The percentage of photons that arrive at a detector and are actually detected is its quantum efficiency. For sensitive low-noise detectors at optical wavelengths, the quantum efficiency is 10–40 per cent. In the near-infrared, the lowest-noise detectors provide a quantum efficiency of 1 per cent or less. Other detectors are available in this region that offer quantum efficiencies up to 80 per cent, but they are a bit noisier. As is explained in Chapter 11, there is also quantum noise from the randomness of small numbers in a measurement of physical parameters. Hence there are parts of the electromagnetic spectrum in which we have achieved good sensitivity, and there are parts in which we have not.

The need for narrow beams

The power required of an interstellar transmitter will be strongly determined by the spread of its beam. In targeting a potentially inhabited star system, the beam should minimize the energy that falls outside the habitable zone. The beamwidth for a given transmission is limited by both the diameter of the antenna and the wavelength used. The shorter the wavelength, the narrower the beam for a given antenna diameter. For a narrow beam to be useful, it must be very accurately pointed. Pointing uncertainty can arise from mispointing and jitter of the antenna (Figure 7.6). It is essential that the maximum disturbance of these effects be significantly less than the beamwidth. At optical wavelengths one can make a beam as narrow as one-tenth of a microradian, where 1 radian is 360 degrees divided by the square of the constant pi, whose value is 3.14, and so just under 57 degrees. To achieve such accuracy is akin to throwing a baseball from the mound at Busch Stadium in St. Louis and having it make a strike of the home plate in Wrigley Field in Chicago, 480 kilometers away. This is within the capability of our technology, so we can safely presume that it must be attainable by any civilization that wishes to signal across interstellar distances. Figure 7.7 shows the relationship between antenna diameter and beamwidth in terms of wavelengths. For a given wavelength, it tells us the size of the antenna required to produce a specific beamwidth.

In addition to the geometric spread of the beam with increasing distance traveled, we must bear in mind that space is not actually a complete vacuum and consider the effects of the interstellar medium. However, the interstellar clouds of gas and dust are most dense in regions of high star density, and there are reasons to believe that this environment is not conducive to the long-term

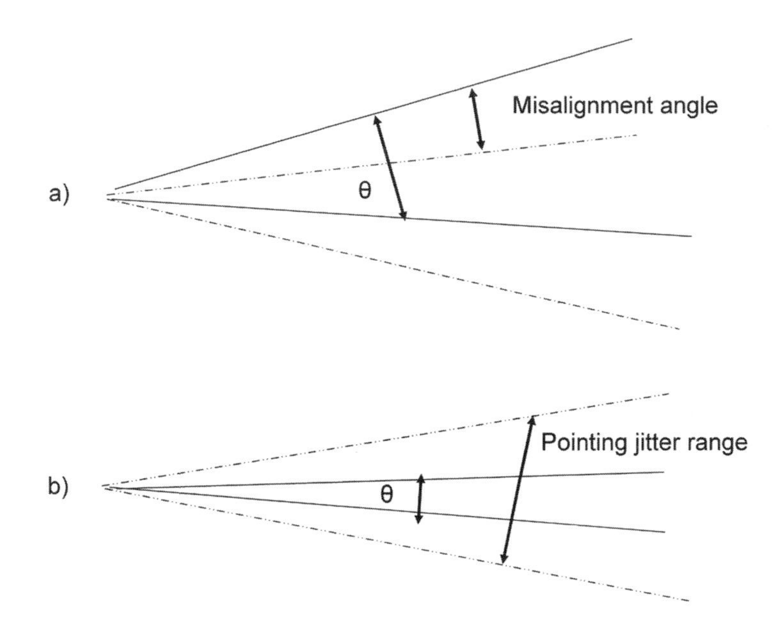

Figure 7.6. Antenna pointing errors: (a) misalignment angle, and (b) pointing jitter.

survival of complex life. In more 'open' areas, the very fact that we can see the stars means that optical wavelengths are a viable option for sending signals.

In textbooks, beamwidth is usually given in terms of antenna gain. This is helpful to engineers in calculating the parameters required to detect a signal and 'close the link', by which we mean achieve an acceptable signal-to-noise ratio. But specifying transmission beamwidth in this way is misleading, since the antenna does not really have gain in the sense of the power emerging from the antenna exceeding that going into it. What antenna gain does, is to send all of the power in a preferred direction so that the measured power will be much more at the angle the antenna is pointing than at any other angle (Figure 7.8). Although it is conventional to refer to this as gain, it would be better to call it a power distribution pattern, since that is what it actually is. The narrower a beam, the greater its power density, which is the power per unit area of the beam.

The actual angle, θ, measured in radians, is more or less the wavelength divided by the diameter of the transmitter in meters. Since the wavelength is a million times smaller in the optical part of the electromagnetic spectrum than for microwaves, the antenna gain far exceeds that of the largest microwave antenna. Even the 305-meter-diameter microwave antenna of the Arecibo Observatory, which is the largest in the world, has a θ-angle measured in milliradians at best. In contrast, an optical telescope 20 cm in diameter is theoretically capable of providing a 10-microradian beam, but in practice the beam can be narrowed only until the pointing uncertainty impairs it. For a θ-angle of 1 microradian it would be desirable for the beam location to be known to within 10 per cent. If the uncertainty is excessive, it may not remain on the

Antenna diameter (meters)	beamwidth (microradians)	wavelength (micrometer)
1	1	1
10	0.1	1
1	10	10
10	1	10
1	100	100
10	1000	10,000
1	100,000	100,000
10	10,000	100,000
1	1,000,000	1,000,000
10	100,000	1,000,000
100	10,000	1,000,000

Figure 7.7. Antenna diameter and beamwidth in terms of wavelength.

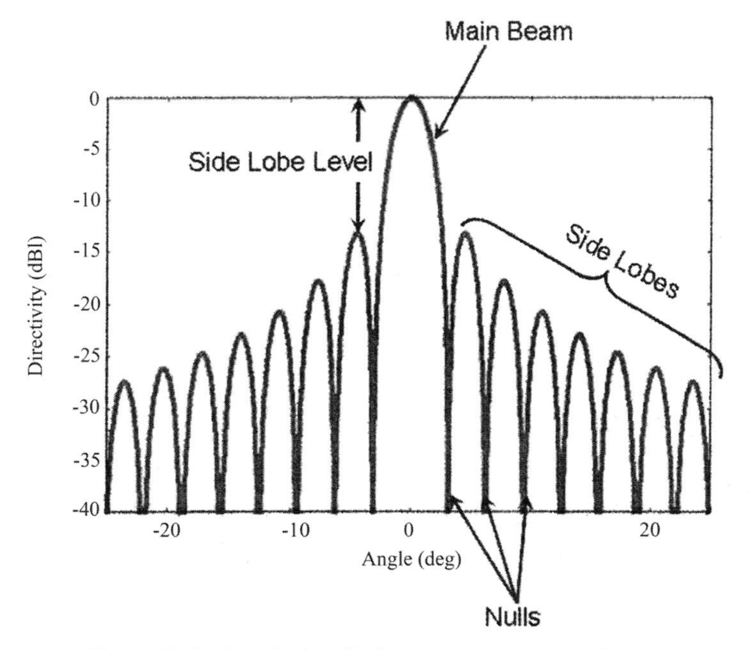

Figure 7.8. A typical radio-frequency antenna gain pattern.

target for the entire transmission. Both static misalignment and dynamic error have to be taken into account in determining pointing uncertainty. Although at times these can tend to cancel out, at other times they can tend to add up. It is necessary to take the worst case into account.

Hence antenna gain is not really gain; the shorter the wavelength, the easier it is to achieve a high antenna gain; and it is important because it increases the power in the preferred direction.

8 Receiver basics and how big is big?

It is fundamental that to detect a signal a receiver is required. A receiver includes an antenna (telescope) of some sort to collect the signal and feed it to a detector which, typically, converts electromagnetic waves into an electrical signal in order to trigger an electronic apparatus by exceeding the random noise of the detection system. The receiver is therefore just one part of a communication link, and we must understand all the key elements of a link before we can delve into aspects over which we have significant control, such as the size of the receiver.

The communication link

A communication system must 'close the link', by which we mean the receiver must detect sufficient signal to recover the information sent by the transmitter. There are four factors common to every link. First, every electromagnetic beam, be it radio-frequency or laser, expands with the square of the distance traveled. The area that a beam encompasses when 10 light-years from its transmitter will be four times that at a distance of 5 light-years, and the signal level per unit area will be one-quarter of that at 5 light-years. The beam spread is given by the angle of transmission, which is approximately the wavelength in meters divided by the transmitter antenna diameter in meters. The beam spread of radio-frequency signals is much wider than that of a laser owing to the difference in wavelength. At a wavelength of 0.1 meter, which is broadly representative of radio frequencies and microwaves, a transmitter antenna 10 meters in diameter would spread its beam over 0.01 radians. In contrast, for a laser at a wavelength of 1 micrometer a transmitter with a diameter of 1 meter would have a beam spread of 0.000001 radian or one microradian. The second factor is the transmission medium. Since we are concerned with such long distances, even small effects can result in large losses to the signal. Particles in the path can absorb or scatter the beam. For example, both radio frequencies and lasers are attenuated in a medium like water. Although an imperfect vacuum, space is generally sufficiently devoid of material that little of the energy issued by a radio-frequency transmitter or laser will be lost to the transmission medium. The third factor is the minimum noise in the receiver. For reliable communication, the signal must exceed this level. If there were no noise, then the minimum possible signal would be a single photon, but the inevitable noise in a receiver means that more than one photon is required for a signal to be detected. In the case of radio frequencies, many photons will be needed since each photon has so little energy compared to the minimum noise. In the case of a laser, the fact that each photon is much more energetic means that it is possible to achieve reliable detection with fewer photons. The uncertainty of photon arrival in a time period may be the only noise. This is often called quantum noise, or noise-in-signal. As related in Chapter 7, the minimum radio-frequency noise is the product of Boltzmann's constant, the absolute temperature of the receiver and the bandwidth; and the minimum noise of a laser is the energy in one photon, which is the product of Planck's constant and the transmitted frequency. We will discuss noise in greater detail in Chapter 9. The fourth factor is collector size. Since the signal is diluted to some level of energy per pulse per unit area, the greater the area of the collector the greater the signal received. The requirements to function efficiently differ for radio frequencies and lasers, but collectors have been developed that use most of the physical area in an efficient way for signal detection. There are some subtler and smaller factors that go into an exact calculation of the link requirements, but they may be omitted herein. The four main factors are interrelated in such a manner that if one knows any three of them it is possible to calculate the other one. So for a given transmitter power and pointing error, if one knows the beam spread, loss to the transmission medium and noise level in the receiver, then the requisite collector size can be calculated.

In attempting to close a link at interstellar distances we find ourselves pushing but not exceeding the limits of today's technology. With further development, our technology will enable us to reliably close the link over distances of many hundreds and likely thousands of light-years.

One can develop a highly sensitive receiver, but without an antenna to collect the signal the system is very limited. What size should an antenna be? This deceptively simple question is at the root of the search for extraterrestrial signals. Antennas are often built of a certain size with a defined collection area for the simple reason that the designer knows the system parameters well enough to calculate the antenna size required to close the link. Usually, in dealing with communication systems such as NASA's Deep Space Network, built primarily to operate space probes, we know the transmitter characteristics. We know how big we need to make the receiving antenna to pull sufficient signal out of the noise. But, if we are ignorant of the transmitter power, beamwidth and wavelength, as we are in the case of SETI, on what basis can we calculate the antenna size? For a large but not impossible transmitter power, our current stock of radio-frequency and microwave antennas are capable of detecting low-bandwidth signals from light-years away. Although the evidence to-date would suggest there is no signal waiting to be found in this part of the spectrum, it could be that the receiver area is still too small to achieve success. The Allen Telescope Array is being built in order to make a much greater effective collection area available to SETI work (Chapter 11).

In the optical regime, antenna size is essentially the area of the primary mirror of the telescope. The new Harvard–Smithsonian system for SETI has a telescope with a diameter of 1.8 meters. Although this is a significant increase over smaller systems, it is well below what is possible using current optical technology (Chapter 12).

If the sender requires the potential recipient of its signal to have attained a certain level of technology, then the power at the target can be tailored so that only a receiving system with a sufficiently large antenna will collect enough photons to detect a signal. Even if someone were looking directly at the signal source, if the area of the collector was insufficient to be sure of collecting even one photon in the measurement time period there will be no detection. (The measurement time period must be short at optical wavelengths, lest background light from the star contribute photons and give rise to a false detection.)

So how big is big for a receiving antenna? For a reasonable presumption about the range of sizes of intelligent beings, the transmitting society could safely assume that any hopeful recipient would build something that was big in relation to themselves, which serves to provide a typical value. This is based on the argument that owing to complex physiological factors including speed of internal neurons, bone structure, etc, the size of an intelligent being is limited by gravity and hence to planet mass. (One paper on the subject showed that for bipedal beings in a gravity field like ours, the maximum size was likely to be 2.4 meters.) It seems reasonable that 'big' would be at least twice and possibly ten times the size of a member of the recipient species. For a being 1 meter tall, this suggests a receiving antenna of between 2 and 10 meters in diameter. For a being at the top end of the size range, the antenna would be between 5 and 25 meters in diameter. Figure 8.1 compares human size to large structures as a means of defining 'bigness'. However, if a society were to pursue projects on a far grander scale then their idea of 'big' might be 100 times our limit, and in presuming that

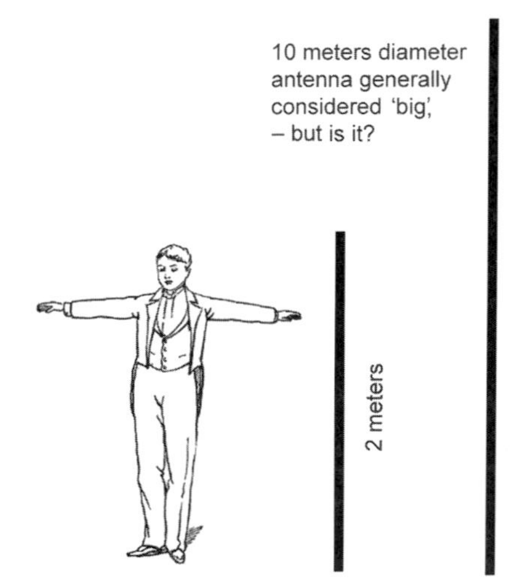

Figure 8.1. How 'big' should we make our receiving antenna? As an example of large-scale engineering, the Pyramids stand 146 meters tall.

others would work on a similar scale such beings might make their transmission so weak as to be undetectable by anyone using smaller, more normal antennas. (As an example, consider how ants co-operate to undertake operations on a scale much larger than the size of an individual.) The lack of success to-date with SETI at radio frequencies and microwaves could therefore merely reflect our small-scale view. It is possible we will require something even larger than the Allen Telescope Array to detect signals that are being transmitted in our direction. So, whilst we do not know the collection area required to detect an extraterrestrial signal, the mantra 'bigger is better' is clearly pertinent.

Although this mantra also applies in the optical regime and we have a number of large telescopes that could be used directly for SETI, the priority is not there. Of the astronomical investigations pursued, the detection of extra-solar planets contributes to SETI by assisting in the selection of targets, irrespective of whether the search is undertaken at radio or optical frequencies. Several new telescopes scheduled to enter service in the coming decade are not only larger, they will also exploit new mirror-stabilization and detector technologies for better performance. The Giant Magellan Telescope (GMT) at Cerro Las Companas in Chile has seven mirrors, each of which is 8.4 meters in diameter. It is not so long ago that a telescope with a single mirror of this size was thought to be enormous. The Atacama Large Millimeter/submillimeter Array (ALMA), located in Chile at Cerro Chajnantor, will use an array of as many as 64 antennas, each 12 meters in diameter, to obtain a collection area of 7,240 square meters. Although the Advanced Technology Solar Telescope (ATST) in Hawaii is a mere 4.24 meters in diameter, it will give a resolution five times better than today's instruments. The Large Synoptic Survey Telescope (LSST) in Cerro Pachon, Chile, is to be 8.4 meters in diameter. The appropriately named Thirty-Meter Telescope (TMT) is to be built in Hawaii, and enter service in 2018. And the European Extremely Large Telescope (E-ELT) will be 42 meters in diameter and made of 984 hexagonal panels. It is to be located either in Chile or the Canary Islands. Construction will start in 2010, and is scheduled to conclude by 2017. It will cost over $1 billion. The image resolution is expected to be at least 10 times better than the Hubble Space Telescope, which has the advantage of observing from above the atmosphere.

This list provides a general picture of the capability of our optical technology, and illustrates how SETI might progress if a large dedicated system were to be funded. It is important to recognize that because SETI is not an imaging task the optics do not require to be state-of-the-art. The laser receiver requires only that any signal in the field of view reaches the detector, that the 'blur circle' at the focus of the optics lie wholly in the detector aperture, and that the focal ratio of the optics give a field of view sufficiently narrow that when the target star is centered there are no other stars to significantly illuminate the detector. And, of course, the collection area should be as large as possible within the budget. It turns out that the optical performance suited to this application is much less costly than for an astronomical telescope. A 0.6-meter-diameter mirror made to SETI requirements will costs many times less than one of the same size which

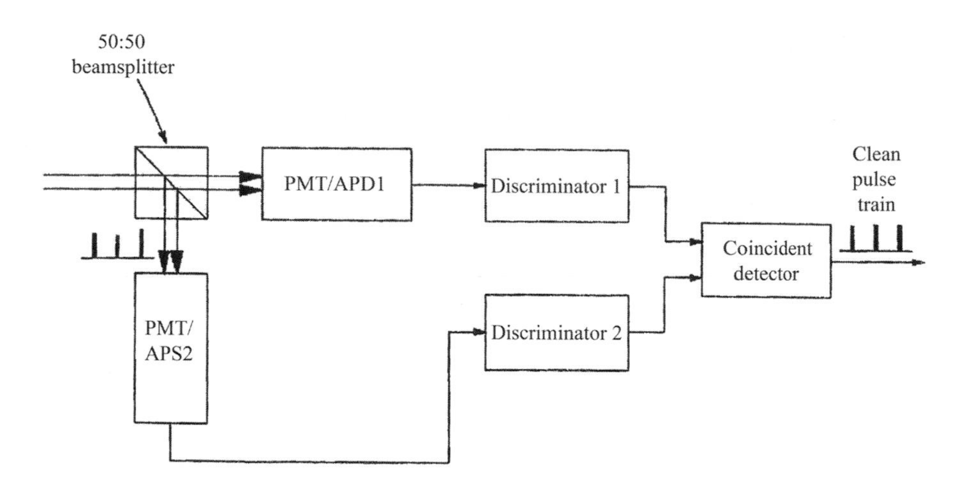

Figure 8.2. Coincident pulses using two detectors to reject false signals.

must produce a pin-sharp image. The light is reflected by the primary mirror to a smaller secondary mirror that sends it to the focal point and the detector assembly, which for SETI is really two parallel channels of photodetection, amplification and threshold detection. The outputs of the threshold detectors are sent to a coincident detector that responds only if both channels 'see' a pulse at the same time (Figure 8.2). This strategy prevents internal noise events in each channel from generating false detections. An optical filter in front of the beam splitter may be used to help to reduce the background light. Although one such system is unlikely to yield results, 10,000 systems distributed around the globe, all aimed at the same time at a particular target star in order to achieve a large effective collector area should be able to detect with great sensitivity any signal from the target star.

Basics of receiving systems

Before we delve into the various SETI approaches, we should review certain basics of receiving systems. One way to increase the chance of detection is to exploit the fact that the transmission occurs at a single frequency called the carrier frequency. The receiver can use a local oscillator to compare its input with the expected signal frequency and so increase the signal-to-noise. At radio frequencies, microwaves and millimeter waves the essential components are an antenna and a local oscillator to convert the incoming wave into a fixed intermediate frequency, and a demodulator designed to extract the carrier and leave the information in the sidebands (Figure 8.3). Amplifiers may be employed at different stages in order to optimize performance. The antenna collects the received energy that arrives from space as a plane wave, by which we simply mean that the waves are traveling in parallel to each other. The waves are focused

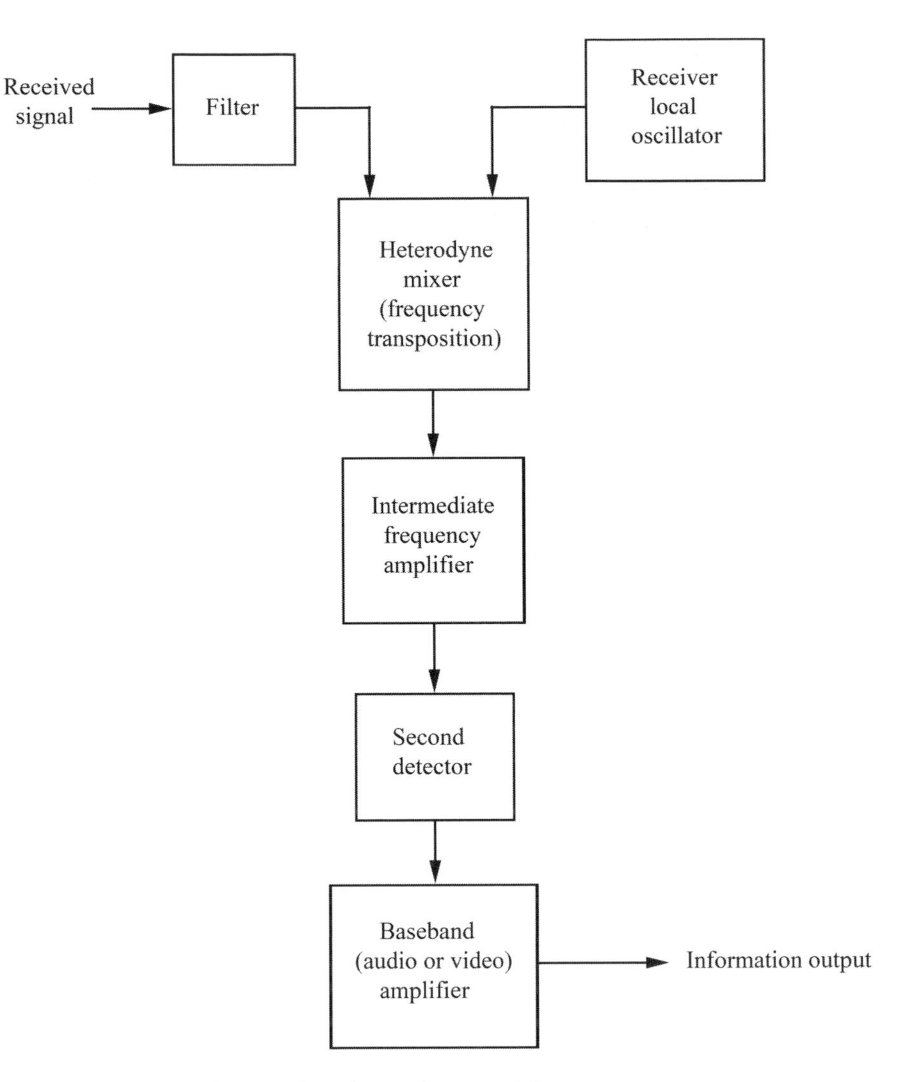

Figure 8.3. A heterodyne receiving system.

by the antenna to form a small spot on the detector. The energy collected by the system is the energy density of the signal multiplied by the antenna collection area. If the incoming wave can stay in phase across the full diameter of the antenna, then the bigger the antenna the better. The waves reflected off the primary dish of the antenna might then be reflected off a secondary or sent directly to the detector apparatus. Deterioration of the signal occurs if all the waves from the antenna do not arrive at the focus at the same time. To reinforce, the waves from different parts of the antenna must be in phase. If they are out of phase they will tend to cancel out. This whole process is called coherent detection, in that the phase of the carrier frequency is a factor in the quality of reception. The bandwidth (i.e. the frequency window or range of frequencies of the intermediate frequency) can be quite narrow, reducing the noise. This technique has been very useful at radio frequencies, but is not as desirable at optical wavelengths.

For non-coherent detection (direct detection) there is no frequency discrimination of the carrier wave to reduce the frequency spectrum that is searched for a signal. No carrier discrimination is undertaken except for the carrier frequency response of the detector itself, which is not very limiting, or in optical systems by placing a filter before the detector to allow only selected wavelengths to pass. Radio in the 1920s used a direct detection crystal system. A great jump in performance came with the introduction of coherent detection (also known as heterodyne detection) and this is used by all radio-frequency systems today.

But as frequency increases, the smoothness required of the antenna increases. At microwaves the wavelength is 3 centimeters and a phase distortion amounting to a small percentage of this can be tolerated. An error of 0.03 centimeter starts to cause degradation, since some of the energy arriving at the focus will be out of phase and will tend to cancel out. For a wavelength of 0.3 centimeter, a flaw of 0.03 centimeter would result in a more serious phase distortion. Hence for millimeter waves a much more accurately figured antenna is required.

In the optical spectrum, coherent and direct detection are both feasible in principle but in practice direct detection is preferred in open-beam systems. There are several major reasons for this. First, in direct detection one need not have prior knowledge of the frequency of the carrier, a photodetector can sense across the spectrum from the ultraviolet to the near-infrared. The carrier is discarded with all the information in the detected signal. No local oscillator is needed. Although it is important to track out Doppler shift in coherent detection systems, it is not an issue in direct detection because the carrier frequency is discarded. Noise in a non-coherent optical system is different from noise in a coherent radio-frequency system, such that narrow pulses which require bandwidth can be achieved without introducing a lot of additional noise. The bandwidth is approximately 1 divided by the pulse width, and therefore a 1-nanosecond pulse requires 1 gigahertz of bandwidth for proper processing. The concept of a direct detection system is quite simple (Figure 8.4).

An analysis based on our best understanding of the issues and a limited amount of observational data suggested 30 good candidates for SETI within 100 light-years of the Sun. Assuming an even distribution of civilizations in space, extrapolation gives 240 candidates within 200 light-years, 1,920 within 400 light-years and some 16,000 within 800 light-years. However, that study focused on the issue of habitable zones. Other factors that reduce the probability of finding an advanced civilization include the fraction of those candidates that develop intelligent life, then those that go on to develop the technology to transmit signals, and finally those that decide to reveal themselves by doing so. If we assign 50 per cent probability to each of these factors this significantly reduces the overall probability. And then there is the lifetime of an advanced

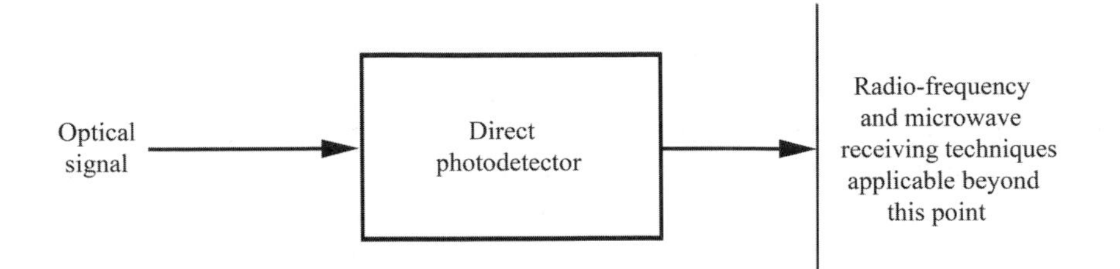

Figure 8.4. A direct detection receiver.

civilization. The range of possibilities for the overlap of civilizations is quite large. If we assign 100 million years as the time period of possible candidates and assume an advanced civilization can exist for 100,000 years of this time, there are 1,000 segments of 100,000 and they have a probability of 0.001 of overlapping in time. Extrapolating further afield to include a lot of new star formations, we find 128,000 candidates within 1,600 light-years. If we take this number and divide first by a factor of 8 to account for the issues above and then divide by 1,000 to account for the improbability of overlap, then we get 15 candidates within 1,600 light-years. It can be argued by analogy with Earth that the time required by multi-cellular life to develop an advanced civilization is about 500 million years. If the lifetime of this civilization is only 50,000 years, the probability of overlap is 0.0001 and there are only 1.5 candidates. Whatever the actual figure is, this line of reasoning implies it is likely to be very small.

Different approaches to large optical collection areas

The signal-to-noise for an optical system can be quite low. The noise due to signal fluctuation is the square root of the number of signal photons that are converted into photoelectrons. If 1,000 signal photoelectrons are detected, the average fluctuation is therefore about 31 and the noise is only about 3 per cent of the average signal value. Although the attraction of a large optical receiver is obvious, it is worth considering whether many smaller units all aimed at the same source might suffice. If we had 10 separate collectors, each received 100 photoelectrons on average, and we combined their detection outputs to obtain a total of 1,000 photoelectrons, then each detector would have a fluctuation of 10 (square root of 100) and the combined noise would be 100 rather than 31. Whilst from a signal-to-noise point of view it is better to use a single collector that is as large as possible, in some cases it could be that combining smaller systems is adequate. Chapter 13 describes in detail how to correlate smaller telescopes to act somewhat like a much larger instrument. It is possible to correlate 1,000 telescopes with 0.6-meter optics that are distributed geographically using a network of PCs, and achieve a timing accuracy of better than 10^{-8} seconds for sophisticated analysis. However, we could be off by an order of magnitude in our grasp of what 'big' means in this context. The Allen Telescope Array, as large as it is, may be 10 times too small. It is quite possible that we simply do not live up to the expectations of an alien transmitting in our direction. In this regard, notice that developing an array rather than a single large facility greatly relieves the constraints of mass, inertia, structural integrity, etc. Arrays offer more scope for achieving very large effective collector areas.

Although SETI at radio frequencies has so far been fruitless, we cannot infer from this that we are alone. It can be argued that aliens are more likely to communicate in the optical part of the spectrum, and it is certainly the case that we have yet to apply our awesome optical capabilities to this endeavor.

9 Noise and limitations on sensitivity

There are two major sources of noise in space. One was discovered in 1965 when Arno Penzias and Robert Wilson at Bell Laboratories in New Jersey were calibrating a microwave communication receiver. They were unable to identify a noise that was present at the same signal strength irrespective of the direction in which the antenna was pointed. They were about to write it off as a footnote in their report when they received a telephone call from Robert Dicke at the University of Princeton, who had heard of their plight. Dicke had calculated that if the universe started in a Big Bang, there should be a background signal in the microwave region, and he suggested this was what the antenna had found. As it happened, the discovery had been made using a radiometer designed by Dicke which compensated system-gain and noise variation by alternating back and forth between the target and a calibrated source of noise. The excess antenna temperature of 3.5K was consistent with Dicke's prediction for a Big Bang. As observations were obtained at other wavelengths, it was established that the power level as a function of wavelength matched thermal blackbody emission at a temperature of 2.75K with its peak at 160.2 gigahertz, equivalent to a wavelength of 1.9 millimeters. In fact, unknown to Dicke, such cosmic background radiation had been predicted in 1948 by George Gamow, Ralph Alpher and Robert Herman. For their discovery, Penzias and Wilson shared a Nobel Prize in 1978. The major source of what amounts to noise at optical wavelengths is the Sun and stars, as the emission of a blackbody with a surface temperature of several thousand degrees has its peak in the visible rather than the microwave region (Figure 9.1).

These two energy sources become very important when looking for signals from extraterrestrials, because they can supply background noise that limits our ability to detect a signal. Of course the Sun can be eliminated by operating at night. However, the star system under observation can contribute background light to the receiver. In subsequent chapters we will see how this occurs, and explain key differences in how noise can be dealt with in different parts of the electromagnetic spectrum. To clarify one point, much of the terminology coined for radio frequencies is counterintuitive in terms of the wider spectrum. In particular, although the terms very high frequency (VHF) and extremely high frequency (EHF) apply to parts of the microwave region they are not really high frequencies compared to lasers. As it would be confusing to talk of lasers using these terms, they will be described in terms of their wavelength instead.

Random variations of energy detected in the receiving apparatus prevent

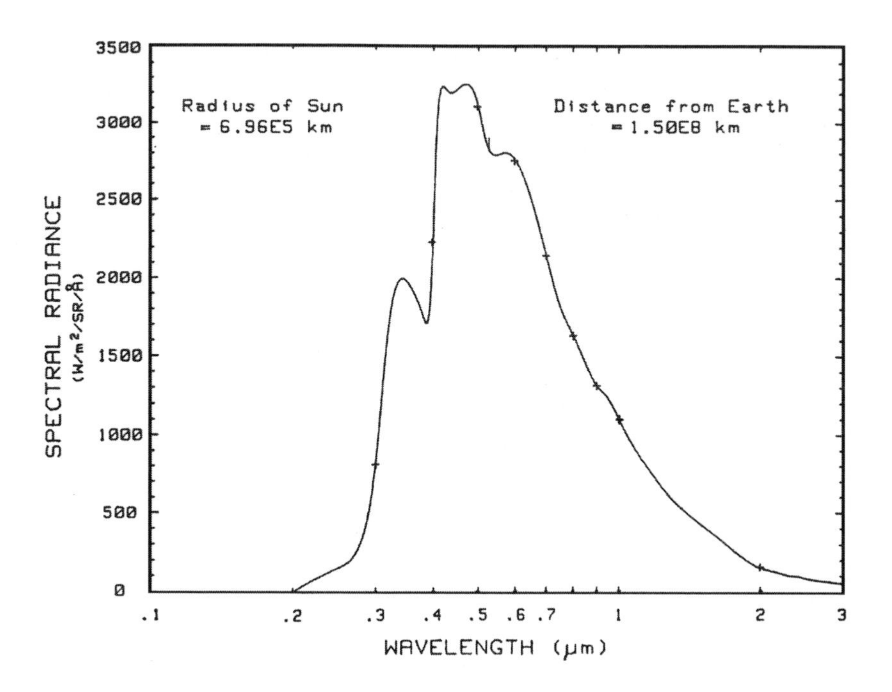

Figure 9.1. Solar spectral radiance.

detection of signals below some value. Every receiver has a highest sensitivity level, and if the signal is less than this level it is said to be 'lost' in the noise. An example is driving a car away from a city that has its own radio station. As the signal diminishes, static creeps in until it is only static (noise). Of course, the signal is still there, it is just too weak to be detected by the receiver. With a larger antenna and a more sophisticated receiver, the distant transmission would be able to be pulled out of the noise.

In searching for signals, why not build the most sophisticated, complex sensitive receiver possible? Even if you are looking in the right direction and are tuned to the right frequency, to detect a signal there must actually be a signal being transmitted towards you and it must be of sufficient strength to be distinguished from the noise. In addition to the environmental noise, there is noise that is intrinsic to the receiving apparatus. Even if a signal is just above the environmental noise, it will not be able to be detected if it is lost in the receiver's internal noise. This is why large antennas are used. For probes in deep space, we know the power of the transmitter and so can calculate the minimum size for the receiving antenna. How large must we make an antenna for SETI in order to ensure that sufficient signal is received to overcome the noise? It is precisely because we do not know how large it should be that SETI has been able to obtain significant observing time on the largest antennas in the world, which were built for radio astronomy and space missions. At radio frequencies and microwaves, therefore, SETI has access to very sophisticated, extremely sensitive, low-noise receivers.

But physics imposes fundamental limitations on receiver sensitivity. As related in Chapter 7, for radio frequencies the minimum noise is the product of Boltzmann's constant, the absolute temperature of the receiver and the reception bandwidth. The giant antennas that stare into space are almost 'perfect' receivers, with the effective temperature of the most sensitive element chilled down to about 4K. The bandwidth for SETI at radio frequencies is very low, typically in the single-hertz regime. This compares to several kilohertz for a telephone or several megahertz for a television. So why not make the bandwidth infinitely small? The answer is that the information rate is directly related to the bandwidth. A signal must change to carry information. If a continuous-wave signal does not change in intensity, phase or frequency, then it carries no information. The act of change introduces the passage of time, and this is inversely proportional to bandwidth. If you increase the rate of signal change (i.e. reduce the time it spends in any given state) in order to generate the ability to send more information, you must also increase the bandwidth. This is why it is difficult to send video over a copper line – since they cannot change the signal rapidly enough, telephone companies slow the video to match the capabilities of the wire, with the result that the resolution is low and a rapidly moving object in the scene generates a multiplicity of images. Optical fiber has much greater bandwidth. The sensitivity of large microwave receivers is therefore near its theoretical limit.

An alternative to building even larger antennas is to increase the effective antenna size by pointing a lot of smaller ones at the same place in the sky and electronically correcting for the phase differences arising from their separation in order to enable them to operate as one large receiver. NASA has such a system, a typical antenna of which is shown in Figure 9.2.

When is "big" big?

Without knowing the characteristics of the transmitter, the best we can hope is that a civilization that decides to transmit probing signals has made a realistic assessment of how large a receiving antenna is likely to be, estimated the environmental noise that would mask the signal in traveling through interstellar space, taken into account the noise intrinsic to a receiving apparatus, and then transmitted sufficient power for the signal to be received. The transmitting civilization may wish to minimize the chance of its signal being detected by anyone other than the intended recipient, and thus will direct the signal at a single star system. Only a laser could achieve this. A radio-frequency or microwave antenna could not be so tightly confined. For example, the target might be 100 light-years from the transmitter and the beam also reaches a star system 200 light-years away, where, having traveled twice as far, the signal strength will be 4 times weaker than that which the sender had calculated to be sufficient for detection by the target. Nevertheless, it may be readily detectable with the receivers available to a civilization in that system, even though it was

Figure 9.2. One of the antennas of NASA's Deep Space Network.

not intended for them. If the signal were to reach a star 500 light-years distant, however, it would be 25 times weaker and a civilization in that system would be able to detect the signal only if it had an antenna much larger than that expected by the sender. We will return to this in a later chapter. Meanwhile, let us further consider the issue of noise.

Noise and its effects on detection

Why is there a random energy variation called noise? It was first studied in the early 1900s by the botanist Robert Brown when observing the seemingly random motions of particles suspended in a fluid. When the mathematics of this 'Brownian motion' proved to be a fundamental phenomenon of physics it was renamed 'thermal noise'. The radio-frequency and microwave noise is of this type, and is best described as blackbody radiation. Thermal noise can be minimized by cooling the receiver, but it cannot be eliminated. Another noise source becomes critical in the optical regime: quantum noise. This results from the fact that each visible-wavelength photon has so much more energy than a microwave photon. As related in Chapter 7, the minimum noise of a laser is the energy of just one photon, which is the product of Planck's constant and the transmitted frequency. There is a difference in frequency of at least 10,000 between microwaves and visible wavelengths, and each optical photon is that much more energetic than a microwave photon. Photons do not enter a receiver in an absolutely steady stream, but with some variation. In every measurement period in which X number of photons is the average, sometimes X will be a little greater and sometimes a little less. The way it works is that if X is the average, then the square root of X is the expected average variation. For microwaves this does not cause any additional noise because, with a photon having so little energy, the minimum signal level to overcome the thermal noise must consist of many thousands of photons. The square root of a large number like 10,000 is only 100, which means that in one case 9,900 photons are detected and in another 10,100 are detected, which is a variation of only plus or minus 1 per cent. But in the optical spectrum thermal noise is small, enabling sensitive detectors to essentially count individual photons. If the number of photons received is small on average, then the percentage variation will be large. If instead of 10,000 microwave photons only 25 optical photons were received, then the square root of 25 is 5, which is 20 per cent of the average and this can result in a measured value between 20 (i.e. 25–5) or 30 (i.e. 25+5). In fact, for a significant part of the time it will fall more than one standard deviation down to 15 or up to 35, but the nominal value remains 25. The difference between thermal noise and quantum noise as a function of wavelength is illustrated in Figure 9.3, which shows the thermal noise dropping at laser wavelengths and the photon noise climbing with frequency to dominate at visible wavelengths.

Now, if thermal noise is not limiting the system, could just one detected photon be sufficient for a signal? Physics say not, owing to the uncertainty of arrival time of a photon. If one is the average, then sometimes two or more will be measured in many measurement intervals and in other intervals zero will be measured. Low numbers of individual events such as photon arrivals are described by Poisson statistics. Figure 9.4 shows Poisson statistics for the probability of detection when the average number of photons measured in an interval is small. Figure 9.5 shows the more general case. This is the classic situation for two distributions, one due to non-signal events alone and the other

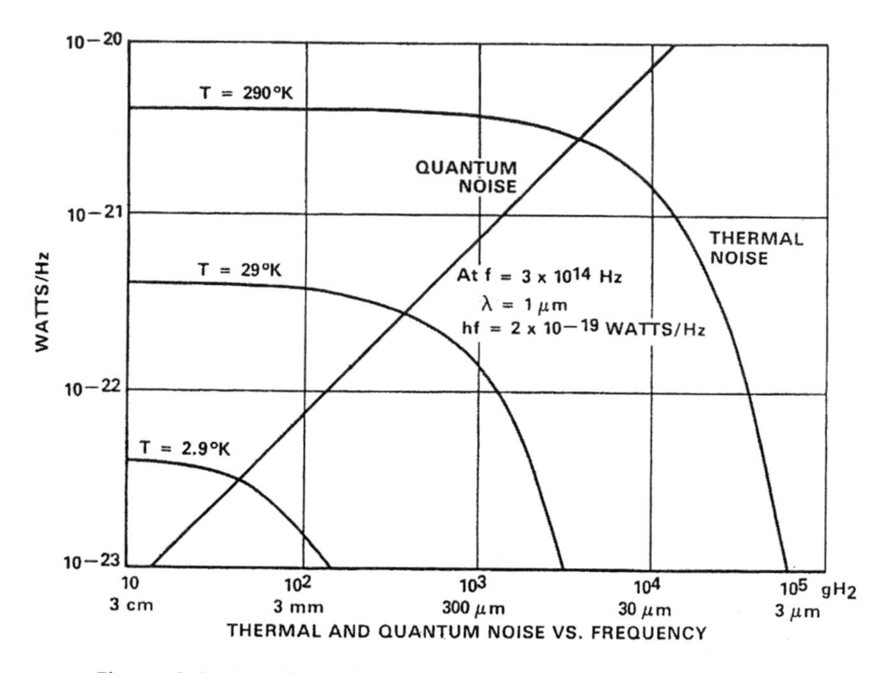

Figure 9.3. How thermal and quantum noise vary with frequency.

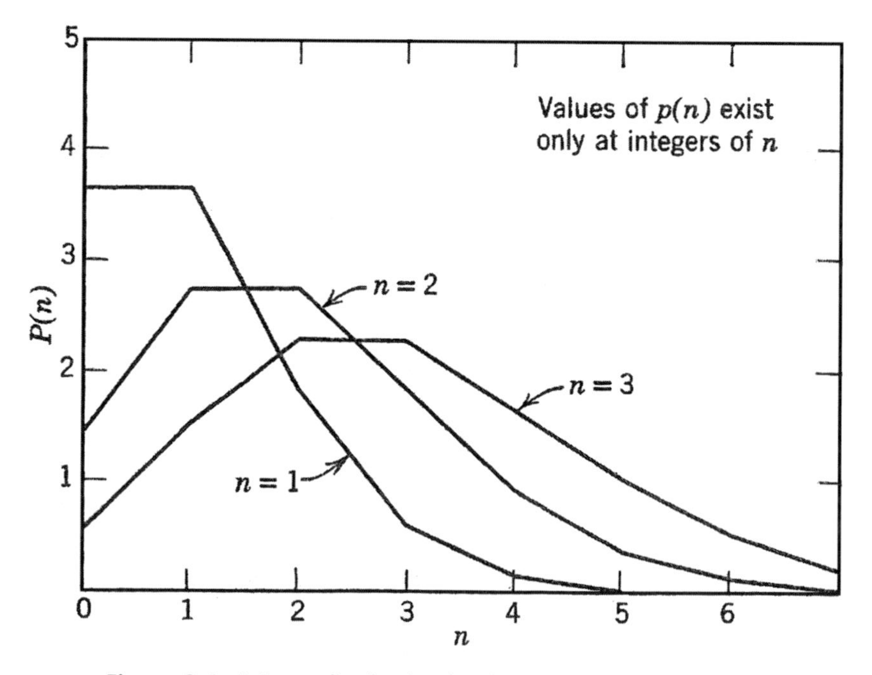

Figure 9.4. Poisson distribution for discrete cases, n = 1, 2 and 3.

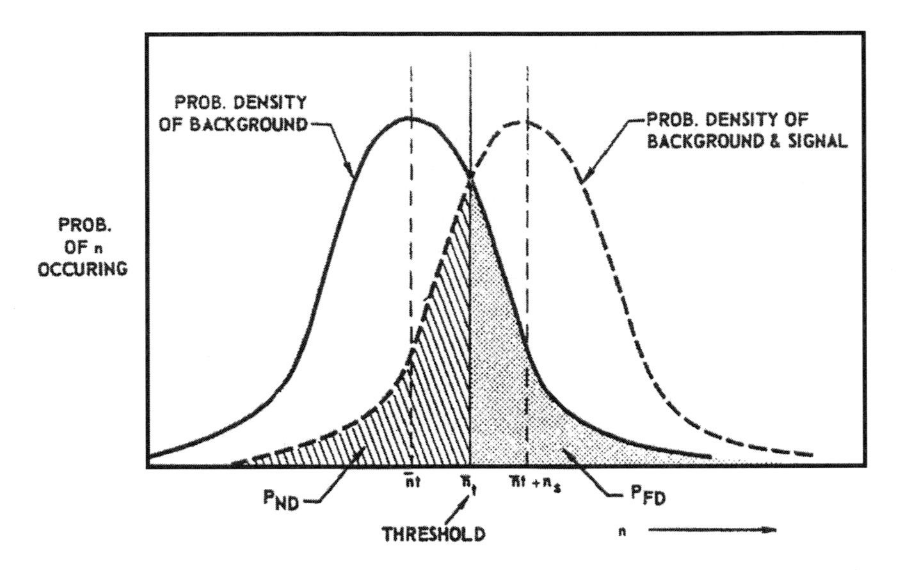

Figure 9.5. Probability distributions of received energy for background (i.e. non-signal) and signal plus background (i.e. signal-added) cases showing the threshold.

caused by the signal adding to the non-signal. The threshold is set between them. The detector cannot separate a signal photon from a non-signal photon. If the difference between signal and non-signal event distributions is sufficiently large to provide a good separation then it becomes easy to decide when a signal detection is made for the reason that it is greater than the threshold. When the distributions are closer because the average signal level is less than the non-signal power, errors can arise from (1) false detections, (2) missing the signal when it is present, and (3) a combination of these depending on where the threshold has been set. Because these distributions derive from the quantum nature of physics, they effectively act as noise by making it difficult to distinguish a signal from non-signal energy. Figure 9.6 shows the signal-present decision when the threshold has been reached in photoelectrons for that short measurement period. In order not to have a large number of false detections, the threshold must be set high, but this requires the signal level to be a certain value in order for it to be detected reliably all the time.

This is the dilemma of the 'noise problem'. At radio frequencies and microwaves the signal must exceed a minimum level if it is to be pulled out of the thermal noise. At optical, infrared and ultraviolet wavelengths, quantum noise limits detection. In both cases thresholds must be set in order to avoid false detections, and these in turn require a given signal level to enable successful detection. At radio frequencies and microwaves the technology has been perfected to such an extent that there is only a factor of two scope for improvement. In optical, infrared and ultraviolet the possible improvement is much greater. Single-photon detectors exist, but they cannot detect every photon. The name means the detector is capable of responding when as few as

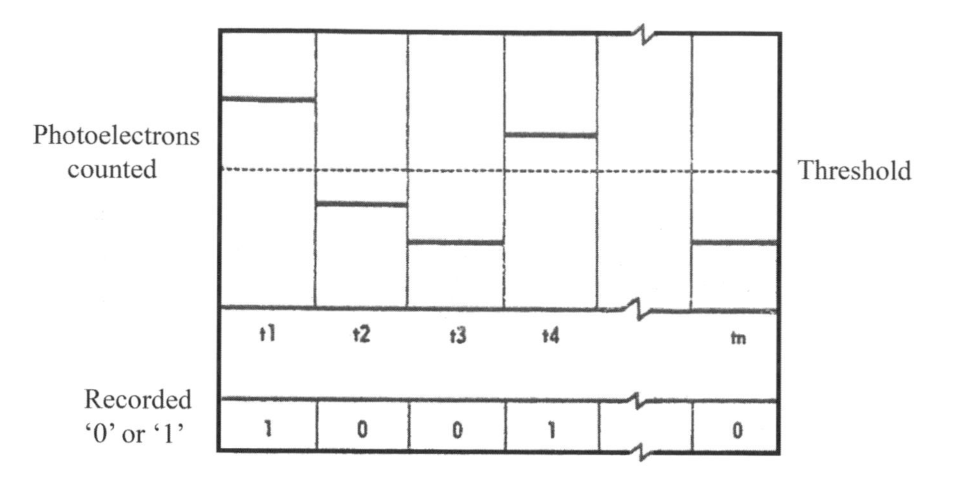

Figure 9.6. Threshold detection. Decision-making receiver waveforms showing the number of photoelectrons counted per measurement time. If the count exceeds the threshold then a '1' is recorded, otherwise a '0' is recorded.

one photon is detected. It does not take into account the parameter called quantum efficiency, which is the percentage of the photons entering the detector that go on to be detected. For different devices at differing wavelengths in the visible and near-infrared, quantum efficiency can be as low as 1 per cent or as high as 80 per cent. A single-photon detector with 50 per cent quantum efficiency actually requires two photons to enter it to register a detection.

One final point regarding the background noise from the targeted star system. The spectrum of a star includes gaps called Fraunhofer lines. These are wavelengths at which energy has been absorbed by the envelope of the star, and so appear dark in the spectrum. Figure 9.7 gives the lines in the visible spectrum of the Sun. A receiver which is tuned to such a wavelength by a narrow-band optical filter will reduce the number of background photons received. In the case of looking at stars far from the Sun, it is necessary to allow for the Doppler shift arising from the differing motions of the stars in space. The complexity of looking at a different wavelength for each target star may make it less desirable to implement a receiver this way. One of the major benefits of using short pulses and direct detection of photons is that the exact wavelength of the signal is not important. The background is sufficiently low that there is no need to use a narrow band in an effort to reduce noise. Direct detection can readily accommodate Doppler shifts of up to 100 gigahertz.

In order to decipher a signal, the type of modulation utilized must also be known. Everyone knows that a radio transmission can be either amplitude modulation (AM) or frequency modulation (FM). The former detects the variation in amplitude of the carrier wave and the latter detects small variations in the transmission frequency. To preclude interference, they are used in different portions of the spectrum. In addition to these analog modulation

3933	CA
3953	CA
4101.750	H
4226.742	CA
4307.914	FG
4307.749	CA
4340.477	H
4861.344	H
5167.510	FE
5167.330	MG
5172.700	MG
5183.621	MG
5269.557	FE
5889.977	NA
5895.944	NA
6562.816	H (Hydrogen Alpha)
6869.955	0
7621	0
7594	0
8228.5	
8990.0	

Figure 9.7. Fraunhofer lines in the visible spectrum in angstroms.

methods, communication systems also use a number of pulse modulation techniques. These include pulse code modulation (PCM), pulse amplitude modulation (PAM), pulse width modulation (PWM) and analog pulse position modulation (PPM). Figure 9.8 illustrates some of these methods. A digital technique called pulse interval modulation (PIM) is useful for optical signals due to its low duty cycle. Radio-frequency communication systems generally use some form of PCM where the duty cycle is typically 50 per cent, and 100 per cent if there is a signal all the time. This method can use one frequency for half of the time and another frequency for the other half, or one polarization for half of the time and an orthogonal polarization for the other half. Noise considerations favor employing a continuous wave for radio frequencies, and employing a low duty cycle for optical wavelengths. When pulses are detected, the modulation has to be determined. This can be done by analysis of the received signal.

PCM is a digital method for transmitting analog data. There are only two possible states, '1' and '0'. This remains true no matter how complex the analog data stream. To obtain PCM from an analog waveform, the analog signal at the transmitter is sampled at regular time intervals (Figure 9.9). The sampling rate (i.e. the number of samples per second) is several times the maximum frequency of the analog signal in hertz. The instantaneous amplitude of the waveform for each sample is rounded off to the nearest predetermined level. This process is

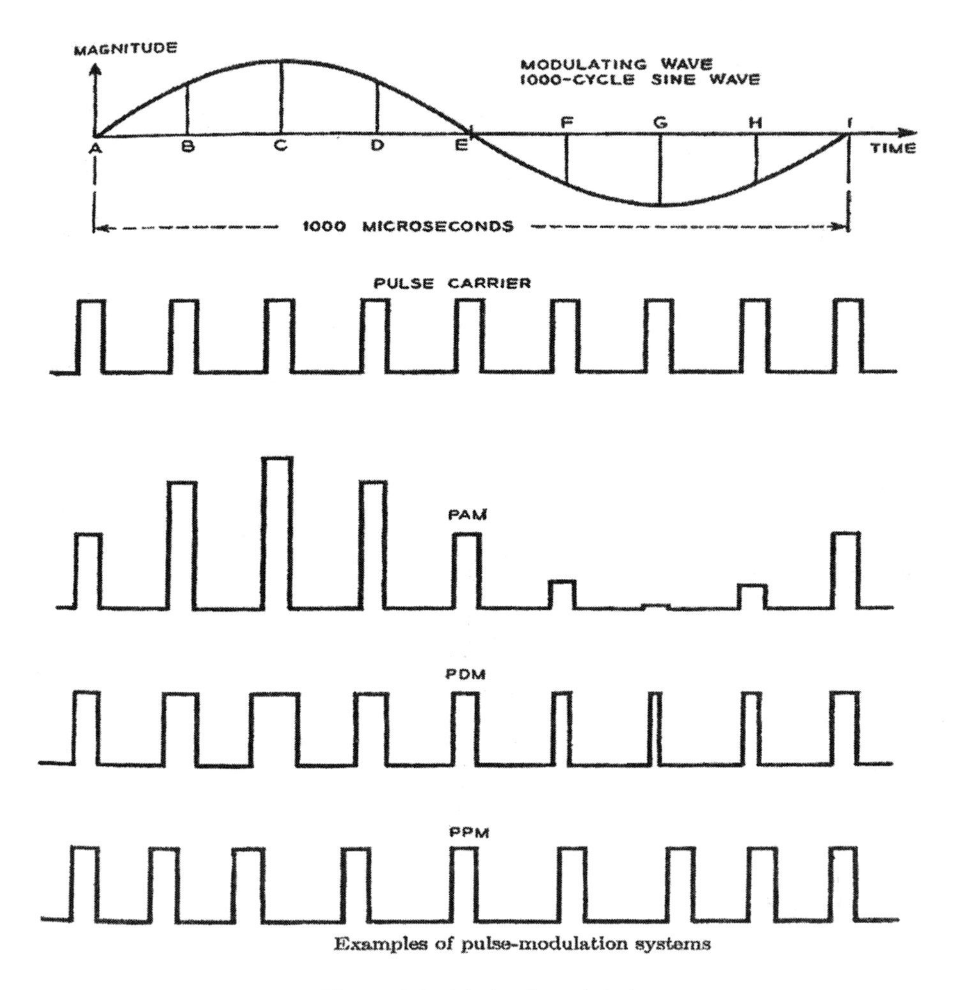

Figure 9.8. Kinds of modulation.

called quantization. The output of a pulse code modulator is a series of numbers. Owing to the use of binary arithmetic, the number of levels must always be a power of 2. The values 8, 16, 32 or 64 can be represented by 3, 4, 5 or 6 bits respectively. At the receiver, this process is reversed. The detected PCM signal is converted by a demodulator and processed to restore the analog waveform. A digital-to-analog converter is used to accomplish this with the necessary accuracy. For each interval in a typical PCM data stream there is a '1' or a '0'. A '1' can be a pulse and a '0' a 'no pulse', or a '1' can be one frequency or polarization of the carrier wave and a '0' the alternate frequency or polarization.

Electromagnetic waves have three important features that can be used to encode information. Although AM and FM are common, polarization is a third possibility. Polarization is the orientation of the electric field of the wave (Figure

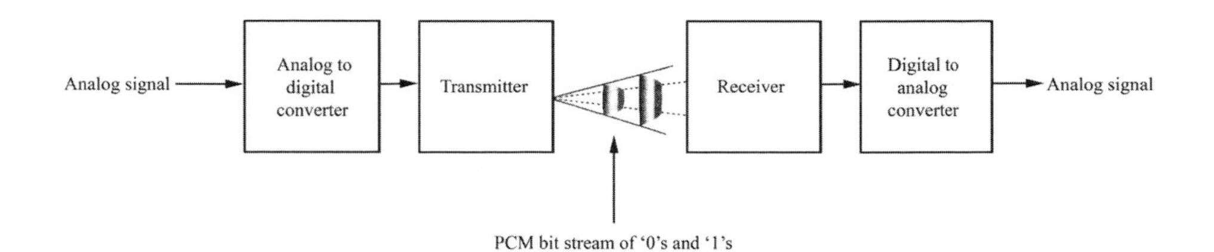

Figure 9.9. Pulse code modulation.

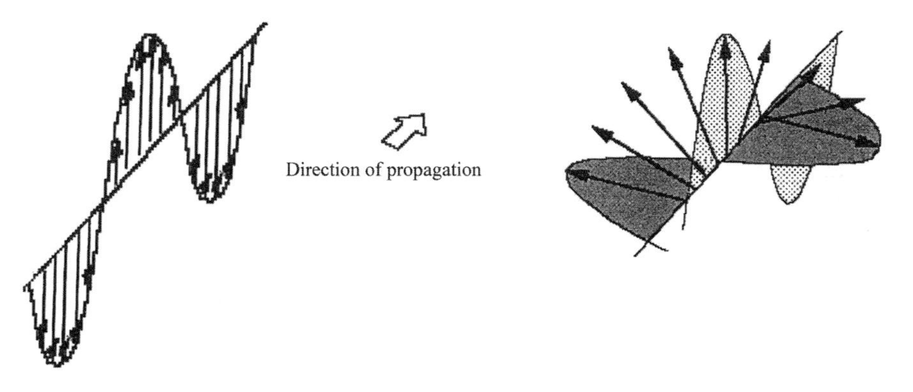

Figure 9.10. Polarization of electromagnetic waves: linear polarization (left), and circular polarization (right).

9.10). When the electric field is oscillating in a single plane this is described as linear polarization. Circular polarization is a combination of two perpendicular linear waves that are 90 degrees out of phase with respect to each other. Circular polarization is used in satellite TV and various military communications. PCM can be sent by changing the polarization in a similar manner to changing the frequency in frequency shift keying (FSK).

There are therefore many ways to modulate, and the choice can be a complex trade off that is beyond the scope of this text. For detection of SETI, it would seem likely that the sender would choose a modulation that is easy to recognize, such as 'pulse' or 'no pulse'. In Chapter 16 we will discuss PIM, which is very attractive in the optical regime.

Part 3: Programs in SETI

10 A brief history of SETI

Only a relatively short time ago, we thought ourselves to be unique in the universe, but the more we find out the less this seems likely to be true. In the 1960s people began to ponder the question: if we are not alone, then where is everyone? A good account of the beginning of the SETI communication work is Frank Drake's 1992 book, *Is Anyone Out There?* The important early papers are in the book *Interstellar Communications*, edited by A.G.W. Cameron and published in 1963. It includes the paper by Giuseppe Cocconi and Philip Morrison in the journal *Nature* in 1959 that first mooted the possibility of surveying nearby stars for microwaves of intelligent origin, and also the seminal paper by Robert Schwartz and Charles Townes in the same journal in 1961 on the possibility of communication by laser. The Cocconi and Morrison paper resulted in activity at radio frequencies seeking interstellar signals. As Drake describes it, Project OZMA was the first organized experiment to look for signals. The selection of radio frequencies was based on the fact that large antennas existed; the suspicion that aliens might opt to transmit at the emission frequency of neutral hydrogen atoms at 1,420 megahertz, which is one of the quietest parts of the spectrum; and because the Earth's atmosphere is transparent at this frequency. What is more, frequencies around that number are protected for radio-astronomy purposes, hence interference signals should be reduced in both number and magnitude. Drake got OZMA initiated in 1960 using the Green Bank antenna of the National Radio Astronomy Observatory. It had limited equipment, especially in receiver sensitivity, compared to what was coming downstream in technology, but it was the start of the search for signals. It began by looking at Epsilon Eridani and Tau Ceti, two nearby solar-type stars. Coconi and Morrison also recommended frequencies in this part of the spectrum. In fact, because the band from 1,420 to 1,640 megahertz lies between the hydrogen spectral line and the strongest hydroxyl spectral line, in 1971 Bernard Oliver nicknamed it the 'water hole' since the lines are the dissociation products of water, which is considered crucial to life.

Space missions that will help

NASA's Kepler mission considers an 'Earth-type' planet to be one with a mass that is between 0.5 and 2.0 times that of Earth, which corresponds to between 0.8 and 1.3 Earth radii. The mission is capable of detecting planets of up to 10 Earth masses, or up to 2.2 Earth radii. According to our best knowledge of how planets are formed, a rocky planet with rather less than 1 Earth mass will not have retained an atmosphere and will be an unlikely candidate for life, and one with over 10 Earth masses would have attracted sufficient hydrogen and helium to become the core of a gas giant. The concept of a habitable zone was first developed for roughly circular planetary orbits, but an eccentricity can cause a planet to spend some time in such a habitable zone. If Earth were to have an eccentricity of 0.3 then its heliocentric distance would vary between 0.7 and 1.3 AU. The energy received in sunlight would vary inversely with the square of the heliocentric distance, but the Earth's speed would be fastest when closest to the Sun and slowest when furthest from the Sun, and the oceans would act to store heat when near the Sun and release it when far from the Sun, with the result that the temperature would remain conducive to life as we know it. At eccentricities greater than 0.3 the situation would be questionable. As we gain hard data on rocky planets in other star systems, we will not only be able to refine the statistics used to calculate the likelihood of there being advanced life around other stars, we will also be able to choose specific SETI targets.

An explosion of activity

By 1962 a number of papers had been published which discussed the possibility of interstellar communication and the detection of alien civilizations. An initial flurry of seminal papers were written by I.S. Shklovsky, Su Shu Huang, Freeman Dyson, Giuseppe Cocconi and Philip Morrison, Frank Drake, Marcel Golay, Ron Bracewell, Bernard Oliver, and Robert Schwartz and Charles Townes. When editing *Interstellar Communication* in 1963, A.G.W. Cameron reprinted 29 papers that he considered of primary importance, all written between 1959 and 1962. This was clearly a period of intense speculation about the possibilities. These papers initiated a number of radio frequency and microwave investigations. In 1973 Ohio State University introduced a football-field-sized radio-telescope. In the 1980s SETI became more serious owing to the vast improvements in low-noise receivers and data processing developed by NASA, and the large antennas built by NASA and others. Paul Horowitz of Harvard University led an effort called Project Sentinel using a 26-meter-diameter radio dish on which time was available, and covered 65,536 closely spaced frequencies around 1,420 megahertz. In 1983 the University of California at Berkeley began a program called SERENDIP, an acronym for Search for Extraterrestrial Radio Emission from Nearby Developed Intelligent Populations. Over the years, this effort has grown in capability. Its receiver was connected to various radio-telescopes around the world, and monitored the incoming data stream for something other than noise. In 1992 the Arecibo Observatory began SERENDIP III using a receiver with 4 million channels. Although the largest antenna in the world, it is built into a hollow in the ground on an island in the Caribbean and therefore has the disadvantage of being restricted to between +2 and +35 degrees of celestial declination. NASA became involved in the early 1980s, scanning the sky using the 34-meter-diameter dishes of its Deep Space Network. When federal funding of SETI was curtailed, private donations sustained the SETI Institute, which in turn funded various programs over the years.

One of the largest programs was Project Phoenix, named for having risen from the ashes of a NASA plan that was killed off by the cancellation of federal funding. It developed a truckload of equipment which traveled to radio-telescopes around the world, including about 5 per cent of the observing time at Arecibo, and examined all stars regardless of their spectral class within 150 light-years, a radius that included around 800 solar-type stars. This 9-year search concluded in 2004, having examined more than 2 billion channels between 1.2 and 3 gigahertz at a resolution in frequency of 0.7 hertz. The SETI Institute's follow-on effort, the development of the Allen Telescope Array, is both more complex and more capable (Chapter 11).

Although the initial SETI efforts of any significance were at radio frequencies and microwaves, Schwartz and Townes discussed the possibilities of optical SETI by the use of continuous-wave lasers. In 1965 I myself pointed out that pulsed laser signals allowed a transmitter to use much less energy, and readily overcame the background of the planet's parent star. Short pulses offered an advantage to the recipient, in that direct detection did not require the exact frequency to be known. The use of pulsed lasers could be a more reasonable choice for interstellar communication than either radio frequencies or continuous-wave lasers. But in the early 1990s Project Cyclops, led by Oliver, made a study that concluded that microwaves were the best choice. It had looked at continuous-wave lasers, but not short-pulse lasers and direct detection. The search for laser signals has therefore been needlessly delayed. The early optical SETI projects looked for continuous-wave lasers, and only briefly inspected several hundred stars. However, Harvard has recently commissioned the first all-sky facility devoted to seeking laser pulses. Optical SETI is discussed in detail in Chapter 12.

11 Radio-frequency and microwave SETI, including the Allen Telescope Array

Since we began to search for alien signals in 1960 we have increased the size of our antennas, increased the sensitivity of our detectors and increased the sophistication of our data processing. Despite the failure to detect a signal at radio frequencies and microwaves, it is the very limitations of the various efforts that enable the search to expand in scope. This is because (as noted in earlier chapters) it is possible that we have found nothing because the originating civilization is using a transmitter power that requires a potential recipient to build an enormous antenna in order to pick up the energy required to pull the signal out of the noise. Even the increase in capability provided by the Allen Telescope Array might not be sufficient. And, of course, there remains the requirement that we point our receiver in the direction of the source as it is transmitting towards us and be in the right part of the electromagnetic spectrum. In this chapter we discuss nine projects which operated in the past or are currently in progress.

Argus

Project Argus is an effort to enlist individual amateur radio-telescopes to search for signals. Paul Shuch led it through the SETI League, which is an independent non-academic operation with a number of programs. Over 100 volunteers equipped with antennas of up to 5 meters in diameter are distributed globally. Figure 11.1 shows a typical setup. Argus has calculated that 5,000 antennas will be required to operate in a coordinated way to enable it to achieve an all-sky capability. The observers use the Internet to find out which part of the sky to aim their antennas at. The sensitivity of each telescope is also an issue, because the noise level is greater than in larger and more sophisticated professional antennas. Although a very low budget affair, Argus deserves applause for exploring parts of the sky not previously investigated.

META

The Mega-channel Extra-Terrestrial Assay (META) was led by Paul Horowitz of Harvard University. It used a 25-meter-diameter antenna owned by Harvard at

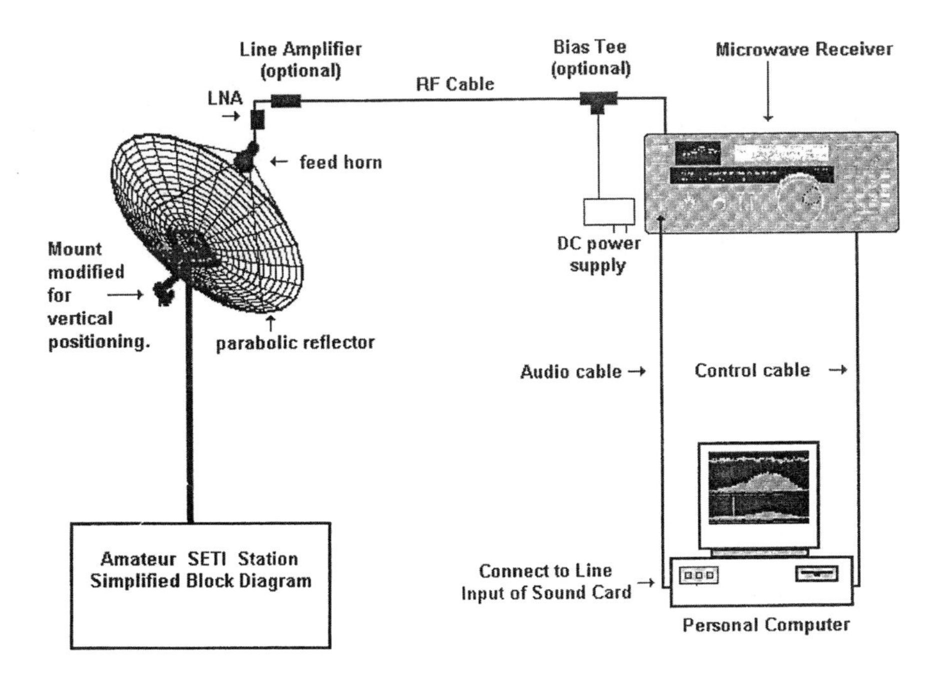

Figure 11.1. A typical amateur radio-telescope setup.

Oak Ridge in Massachusetts, and sought narrowband signals at both the 1,420 megahertz line and its second harmonic using an 8.4-million-channel receiver that comprised a Fourier spectrometer with 400 kilohertz of instantaneous bandwidth and a resolution of 0.05 hertz. In order to avoid false detections from interference sources, it operated in meridian-transit mode, with each potential source passing through the antenna's beam pattern in just 2 minutes. It covered celestial declinations –30 to +60 degrees. META II, was built in Argentina in 1990 using a pair of 9-meter-diameter dishes to extend the coverage deeper into the southern sky. Although SETI researchers in the southern hemisphere can see 100 million stars per square arc-minute in the direction of the center of the galaxy, such a high star density (as discussed earlier) will tend to expose life to a variety of catastrophes that could prevent the development of an intelligent species, so the sheer number of stars present in the southern sky does not in itself increase the chance of our detecting a signal.

BETA

The Billion-channel Extraterrestrial Assay (BETA) was an upgrade and replacement for META. It had 250 million simultaneous 0.5-hertz channels, and the receiver had an instantaneous bandwidth of 40 megahertz to enable three antenna beams to scan the 'water hole' in eight hops of 2 seconds each. During

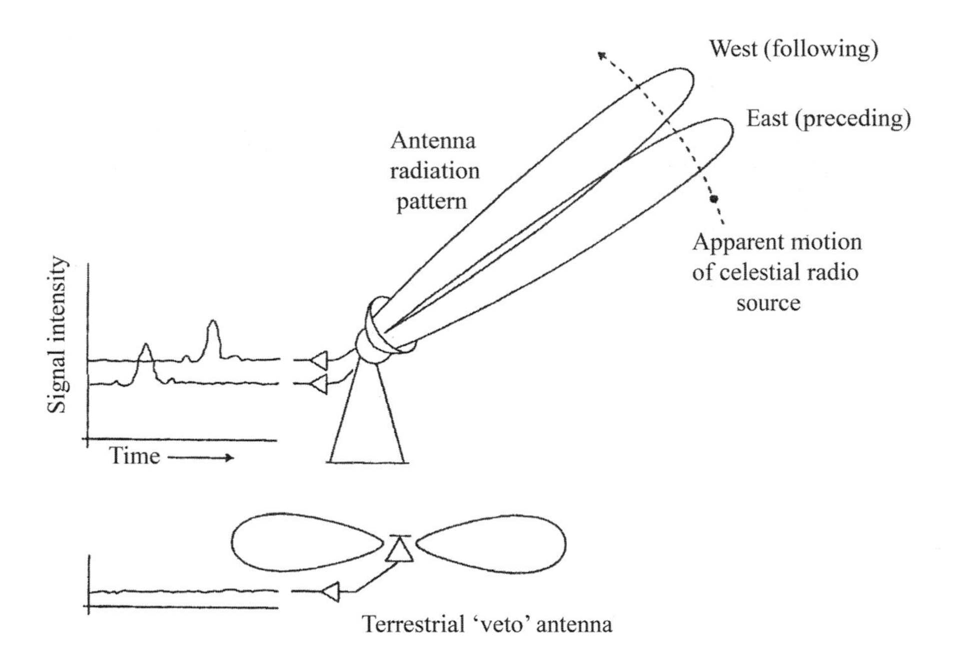

Figure 11.2. The three-beam feed system of Project BETA.

the sidereal drift of a point source across the 0.5-degree-wide beam, the 16-second scanning cycle was repeated eight times. The antenna design exploited the fact that an extraterrestrial signal will have one important feature that terrestrial sources cannot: it will always appear to be coming from a fixed point in the sky, irrespective of the Earth's motion. The 3-beam system used this to eliminate terrestrial interference that could otherwise cause false detections (Figure 11.2). Two adjacent beams were pointed at the sky, one slightly east of the target star and the other slightly west. A signal from a fixed sidereal position would trace out a characteristic pattern of beam lobes, first in the eastern beam and then in the western beam. The third antenna beam provided an azimuth omnidirectional pattern that looked predominantly towards the horizon to pick out interference from terrestrial sources. BETA entered service in 1995, but when winds destroyed one of the antennas in 1999 the project was wrapped up. It covered declinations -30 to +60 degrees with one significant result: no detected signal was classified as 'unknown'.

Phoenix

When Congress cancelled NASA's funding for SETI in 1993, the privately funded SETI Institute was able in 1995 to obtain some of the equipment that NASA had purchased, and then use it for Project Phoenix. This examined nearby stars in greater depth than any other SETI program. It included simultaneous observations by the 305-meter-diameter antenna at Arecibo in Puerto Rico and the 75-meter antenna at Jodrell Bank in England. The system was able to reject false signals by exploiting the fact that terrestrial interference could not simultaneously affect two antennas so far apart. It scanned each target from 1,200 to 3,000 megahertz, but examined only about 800 stars.

SERENDIP

The SERENDIP concept was to piggy-back a special receiver onto a radio-telescope without imposing on the telescope's program, and it has been operating in one form or another since 1978. The current form is SERENDIP IV, and is located at Arecibo. Although it is restricted to the direction in which the astronomers aim the dish, it is able to run more or less continuously, with the result that since 1998 SERENDIP IV has listened to 168 million 0.6-hertz channels in a 100-megahertz band centered on 1,420 megahertz. On the other hand, the Arecibo dish is restricted to a narrow strip of sky, and as the SETI effort is parasitic it cannot match Project Phoenix's ability to examine each star in depth. SERENDIP V, will span a frequency range of 300 megahertz.

SETI@home

SETI took an interesting and imaginative step in 1999 with the SETI@home project. This directly involved the public by making use of thousands of personal computers. In effect, by catching the imagination of the media and the public, SETI gained the equivalent of millions of dollars of free advertising. The essence of the problem was the greater the data the telescopes were able to draw in, the greater was the computer power required to process it. In fact, to fully examine in an appropriate manner the incoming narrowband streams of data required even more computing power than a supercomputer. In 1994 computer scientist David Gedye recognized that 'distributed computing' offered a solution. All that was required was to break the analysis into a number of simple tasks that could be farmed out to PCs to process at times when they would otherwise be idle. An individual who wished to participate in the scheme downloaded a program that fetched 354 kilobyte files (one work unit) recorded by the SERENDIP system. When the analysis was done, the computer sent the results and fetched another work unit. The amount of processing per work unit depended on the capability of the PC, and ranged from a few hours to ten times that. Each work unit analyzed one narrowband segment of 10 kilohertz obtained during 107 seconds on the Arecibo dish, and since the beam was only 0.1 degree wide this represented a tiny strip of sky. It took 250 work units of adjacent data recordings to cover the full 100-megahertz band obtained during this interval. Work units were sent to multiple PCs to eliminate bad results. The value of SETI@home to the SERENDIP project was that it increased the sensitivity 10 fold, enabling the survey to probe deeper into space. With SETI@home handling the data load, 30 times more stars were able to be searched in a given time period.

A system called Berkeley Open Infrastructure for Network Computing (BOINC) was created to give SETI@home greater flexibility. In addition, SETI@home set up a project called AstroPulse to search existing SETI@home data for extremely brief, wideband radio pulses. SETI@home is also extending to the southern hemisphere to process data from a piggy-back detector on the 64-meter-diameter radio-telescope at Parkes in Australia, with BOINC distributing the data. BOINC will also be applied to the new SERENDIP V data stream. The basic point remains, however, the scope for success depends on the choice of 1,420 megahertz, and notwithstanding the vast amount of analysis by SETI@home, if this frequency is wrong then the effort must fail.

Southern SERENDIP

In 1998 a copy of the SERENDIP III receiver was piggy-backed onto the 64-meter radio-telescope of the Parkes Observatory. The receiver had 8.4 million narrowband channels, but in 1999 it was upgraded for 58.8 million channels, each 0.6 hertz wide for a total bandwidth of 35 megahertz. Although the largest dish in the southern hemisphere, Parkes is only one-fifth the diameter of Arecibo and therefore requires a larger signal to overcome the noise. Also, like the other SERENDIP efforts, as a parasite it does not have a capability to evaluate any promising or unknown signal in real-time.

AstroPulse

The AstroPulse project is the first SETI effort to look for microsecond-timescale pulses at radio frequencies. It will detect pulse widths ranging from 1 millisecond to as brief as 1 microsecond. (This is still 1,000 times longer than the 1 nanosecond pulses for the optical regime.) Pulses of 1-microsecond duration might be caused by natural events such as evaporating black holes, gamma-ray bursters, supernovas and pulsars, hence would not necessarily imply an alien signaling. The AstroPulse project will examine archived SETI@home data. This will require an enormous amount of computing power. The analysis must take into account the fact that pulses at radio frequencies suffer dispersion arising from their photons interacting with plasma in the interstellar medium, stretching them out in time. The effect is to delay the arrival of the lower frequencies, turning a sharp pulse into a whistle. Since the phase of the signal is preserved and is subject to compensation at the receiver, dispersion is a coherent effect, but it is complex to shift each frequency component to reinstate the original pulse width. It has been estimated that 0.5 teraflops of computing power is required to analyze a 2.5-megahertz bandwidth. As for SETI@home, the public has been asked to make its PCs available to handle the immense calculations. Pulses in the optical regime will not suffer dispersion, and phase is neither preserved or required to be preserved for high sensitivity because the method of detection (direct) is different.

Small-scale radio-frequency SETI efforts

In addition to the large-scale SETI efforts, which are mainly carried out by Harvard University and the University of California at Berkeley and led by Paul Horowitz and Dan Wertheimer respectively, there have been small-scale projects by amateurs. Paul Shuch of the SETI League took the lead by sponsoring hundreds of amateurs with small radio-telescopes for Project Argus. Although of lesser sensitivity than a professional installation, Argus, with over 100 members, could examine a great deal of the sky.

Other amateur efforts include Project BAMBI, which stands for Bob and Mike's Big Investment. It is a pair of small radio-telescopes in California and Colorado that search the sky in 3.1 million channels. Undaunted by the task, they upgraded their antennas to 4-meter size several years ago. There are also amateur SETI efforts in Australia. The Boonah SETI Observatory, which is currently under construction, will use a pair of 12-meter dishes.

Although one must admire the amateur zeal, success is a long shot with issues of the right frequency, antenna size, limited bandwidth, limited sensitivity and difficulties of determining false detections in real-time. And then, of course, there is the basic requirement to be looking in the right direction at the right time.

Allen Telescope Array

The Allen Telescope Array is a joint endeavor by the SETI Institute and the Radio Astronomy Laboratory of the University of California at Berkeley. The driving force behind this complex project is Seth Shostak, senior astronomer at the SETI Institute. It is being built at the Hat Creek Radio Observatory in California, and when finished it will be an array of 350 antennas, each of which is a 6.1-meter-diameter Gregorian offset primary with a 2.4-meter-diameter secondary in order to minimize sidelobes. At 10,000 square meters, the total physical collection area will be much greater than any other SETI facility. The front end is to be a single cryogenically cooled indium-phosphide Monolithic Microwave Integrated Circuit low-noise amplifier. The array has four main conceptual systems: (1) the antenna collects the radiation from space; (2) the signal path conveys the radiation from the feed located at the antenna's focus to the user; (3) the monitor and command systems allow the dishes to be accurately controlled; and (4) the overall antenna configuration, plus additional infrastructure. Figure 11-3 shows the overall concept and architecture. Figure 11-4 shows the antenna. Figure 11.5

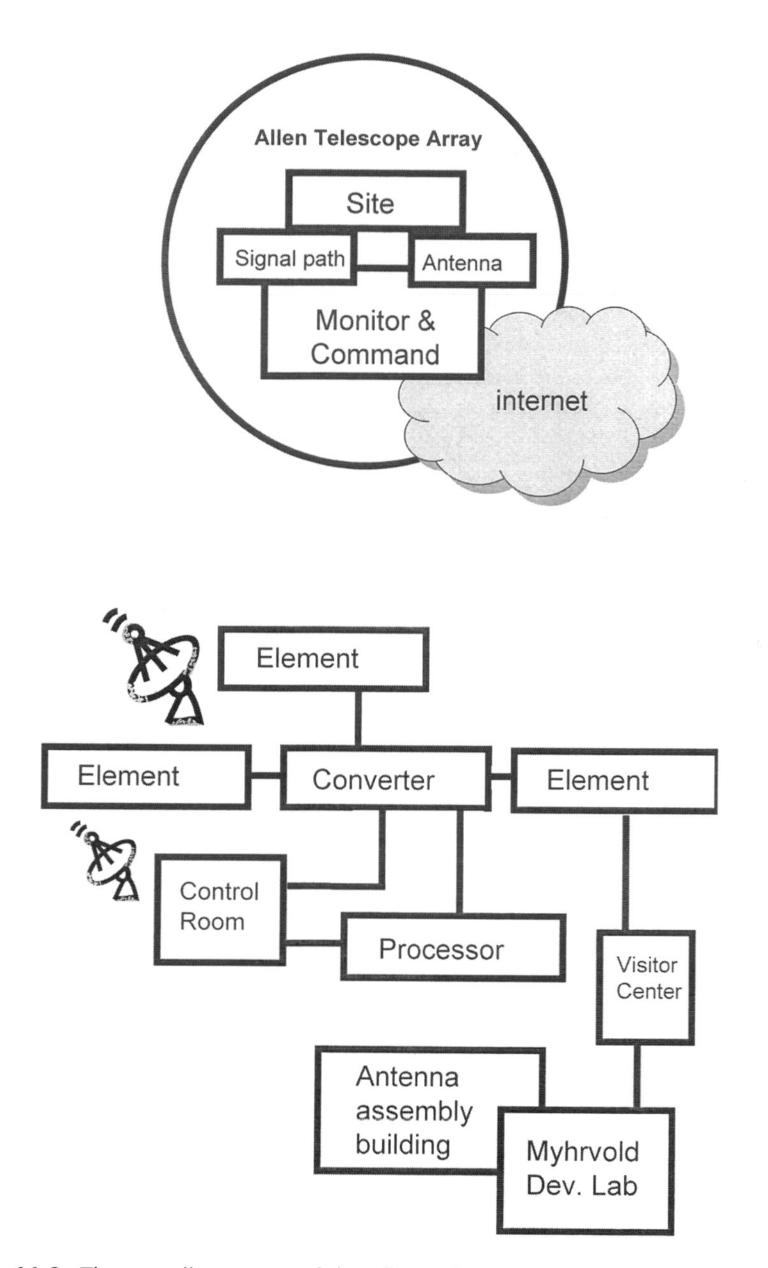

Figure 11.3. The overall structure of the Allen Telescope Array: major systems (top), and physical structures (bottom).

Figure 11.4. Some of the antennas of the Allen Telescope Array.

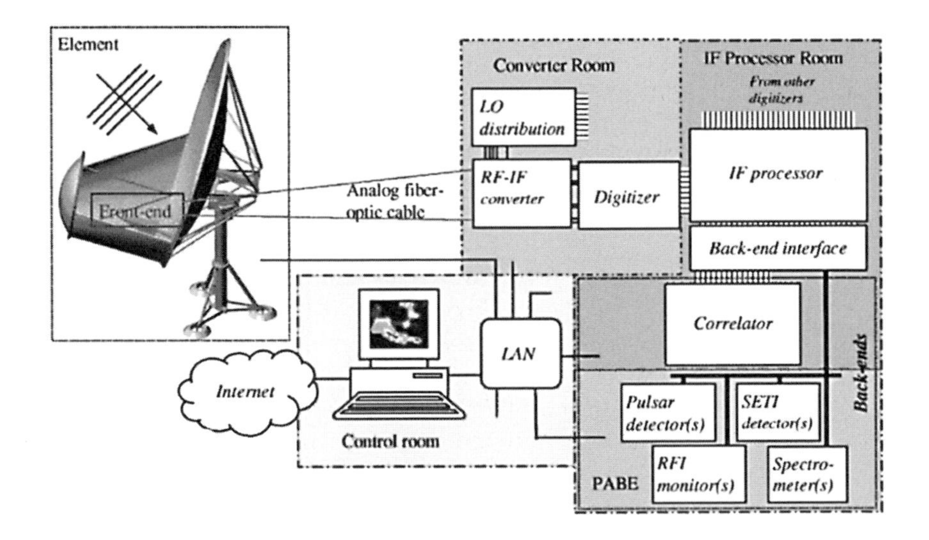

Figure 11.5. Block diagram of the signal path through the Allen Telescope Array.

Array	Beam Size (arc-sec)	key SETI efforts
ATA-42	245 x 118	survey galactic center
ATA-98	120 x 80	targeted survey -100 stars
ATA-206	75 x 65	extensive targeted surveys
ATA-350	77 x 66	more targeted surveys

Figure 11.6. The phased growth of the Allen Telescope Array. The initial configuration (ATA-42) has 42 antennas. The full system will have 350 antennas. Note that 1 second of arc is about 5 microradians.

shows the signal path from sky to user. The array will allow users remote access via the Internet. There are approximately 700 paths from the 350 antennas. The analog fiber-links carry both polarizations of the full instantaneous frequency coverage of 0.5 to 11.0 gigahertz to the radio-frequency converter, which produces four 1-gigahertz-wide dual-polarization channels. These are digitized and sent to the processor. Each of as many as four independent channels is further split into four synthesized beams which can be either pointed at the same part of the sky or steered independently to different parts of the sky. Outputs are sent to either a correlator for combination to make an image or to one of several phased-array back-ends. Some of the back-ends will be unrelated to SETI research, and may include pulsar processed astronomical spectrometers and radio-frequency interference monitors. Low-noise amplifiers are housed in a small cryogenic dewar placed directly behind the antenna terminals to yield a low receiver temperature. The linear dimension of the feed allow an operating range from about 500 megahertz to 15 gigahertz.

The project is known as the Allen Telescope Array because it was initiated by a grant from Paul Allen, cofounder of the Microsoft Corporation. The planned buildup of elements will enable the initial capability to be improved as more elements come on stream. Figure 11.6 shows the planned build up. The first phase was completed in 2007, with 42 antennas. It uses a lot of advanced technology, such as the wideband fiber-link, antenna design and low-noise detector. The pace of build-up is dependent upon funding, but it is hoped to complete the array by the end of the decade. It may, just possibly, have sufficient collection area to pull an alien signal out of the noise – presuming of course that we are justified in searching in the radio spectrum.

12 Early optical SETI and the all-sky Harvard system

An argument can be made that we are more likely to find alien signals in the optical spectrum than at either radio frequencies or microwaves. For one thing, it is easier to deal with noise at optical wavelengths. Attempting SETI at radio frequencies means contending not only with interference from terrestrial sources such as radars and radio stations but also with cosmic sources, including the background left over from the Big Bang. And, of course, there is noise intrinsic to the receiver itself. Although the most sensitive detectors are cooled to almost absolute zero to minimize internal noise, this cannot be eliminated entirely. The only significant terrestrial source of interference for optical SETI is lightning, which is at worst a sporadic problem with a very low probability. In the early days, many investigators dismissed optical SETI, believing that the sender's star would be an overwhelming source of noise. But they did not appreciate that if a short-pulse laser is used instead of a continuous one then it is possible to outshine a star during the time the pulse transmitter is 'on'. With a short-pulse laser, both spectral and temporal discrimination in the receiver can be readily attained since a laser shines at a single wavelength whereas a star shines in a broad spectrum, which enables the laser receiver to reject much of the spectrum but still have a much wider spectral acceptance than microwave receivers. The laser can deliver very large peak powers for brief intervals. And because direct detection for SETI at optical wavelengths is not obliged to preserve the coherence of the signal at the detector, large collectors can be made more cheaply than those of the same size which must preserve the phase of the signal across the face of the collector in order to produce an image.

Photomixing and coincident detectors

It has sometimes been presumed that optical SETI will employ the same method of detection as for radio frequencies, namely heterodyne or coherent detection, whereas it is actually better to use direct detection in which the phase of the carrier frequency is discarded and only the amplitude is detected. Coherent detection is essential for an application such as high-resolution spectroscopy, however, because a laser is used as a local oscillator and mixed with the signal to create a microwave frequency which is analyzed in the same manner as a radio-telescope signal. Owing to the differences in noise at radio frequencies and at optical wavelengths, direct detection can achieve a sensitivity approaching that of heterodyne detection without the complexities and limitations of coherent detection. The type of noise which limits an optical system is the uncertainty inherent in detecting the signal, which is denoted as 'quantum noise' or 'photon noise' (Chapter 9). The major external source of noise is the light of the target star, but this is minimized by seeking nanosecond-like pulses which provide a high peak power at the source without imposing an excessive average power rating on the laser. We have produced picosecond-type pulses with a peak power as high as a petawatt, which is 10^{15} watts. We have also made lasers with several megawatts of average power. Hence it is reasonable to expect that an alien civilization will be able to produce laser pulses with high-peak-power brief pulses and a reasonable pulse rate. The number of photons received in a short pulse can far outshine the natural light from the target star during the time of a pulse. In its simplest form, direct detection consists of a telescope with a photodetector. If a second photodetector is added for coincident detection and the measurement time is short then it will prevent internal noise from triggering false detections. The difficulties of coherent detection include knowing the signal frequency and handling Doppler shifts, but for direct detection a photodetector such as the photocathode of a photomultiplier has a spectral response broader than any Doppler shift likely to be encountered in SETI. The 1-gigahertz shift resulting from a radial velocity of 1 km per second would have no impact on a photodetector, but a heterodyne detector would have to perform the difficult task of tracking this shift and eliminating it. Also, whereas the interstellar medium causes radio frequencies to suffer dispersion, this effect is negligible at optical wavelengths and eliminates the requirement to reconstruct the signal. Although starlight suffers interference in the Earth's atmosphere, this is minor on the 1-nanosecond time scale.

Hence, a 1-meter-diameter telescope will get 200,000 photons per second from a solar-type star at a distance of 1,000 light-years, and for a quantum efficiency of 0.2 over the visible spectrum this equates to 4×10^{-5} background photoelectrons in each nanosecond. As low-probability random events, the arrival of the photons will be in accordance with Poisson statistics. With a threshold set at five or so photoelectrons, the likelihood of receiving five background photoelectrons in the same nanosecond with such a low background rate is infinitesimal. The most likely source of noise is internal to the detector, which is why the receiving beam is split into two paths and a second photodetector added for coincident detection. We can set the threshold at two photoelectrons in each detector. The likelihood of two noise pulses occurring in the same nanosecond is given by:

$$R = (r_1^2\, t)\,(r_2^2\, t)\, t$$

where r_1 and r_2 are the arrival rates and t is the coincident window. If the rates are 10^5 photoelectrons per second and t is 1 nanosecond, then the rate of false detection from the background is 10^{-7} per second, or a little less than once per year. If desired, this can be reduced much further by adding a third detector. The point is that with so few background photoelectrons arriving in each nanosecond, there is an exceedingly small likelihood of there being two events in each of two detectors during the same nanosecond. To express this another way, consider that since there are many fewer photoelectrons per second than there are nanoseconds in a second, the likelihood of a random event occurring in a particular nanosecond is quite low, and the odds of a second random event occurring during the same nanosecond are extremely low. Of course, the larger the collection area of any single telescope the greater will be the background rate, and at some point a third detector may be needed to eliminate false signals. If the same area were achieved by arraying smaller telescopes, then the need for a third detector is reduced. A signal level of five to ten photoelectrons per pulse would avoid false detections and yet guarantee that a signal detection would not be missed. This raises the issue of the receiver's size. All we can do is set a limit. The example was for a solar-type star at a range of 1,000 light-years, which (for reasons given in earlier chapters) is the maximum reasonable distance for SETI. Although stars which are closer will contribute more light to the background, for a given laser power the reduced distance will allow more signal to be detected. This would not be so for more luminous stars such as spectral classes O, B and A, but these appear to be inconsistent with the requirements for the development of advanced life. We can therefore be reasonably certain that light from the parent star of a civilization will not impede our detection of a short-pulse laser signal.

The development of laser communication for submarines, aircraft and military satellites proved that short bursts of laser light are far more efficient than continuous waves at carrying information. Although each pulse has a high peak power, the laser is inactive for most of the time and therefore has a low total power consumption. It is reasonable to believe that an alien civilization must have figured this out as well. With transmissions in brief bursts, each pulse could readily outshine any star in the field of view of the collector. Whatismore, the shorter the pulse, the less background light there is per pulse to compete with the signal. Reducing the pulse to nanosecond intervals makes any signal detected even more obviously of artificial origin, as such short flashes are unlikely to occur naturally. (SETI faced this dilemma in 1967 when a radio source was discovered to be 'ticking' with the regularity of an atomic clock. Until it was realized to be a rapidly rotating neutron star, and natural, the signal was labeled LGM-1, with the acronym standing for Little Green Men.) There is little to be gained from reducing the duration of the laser pulse below 1 nanosecond, because (1) electronics and detectors function well down to nanosecond levels, but they have difficulty at much shorter times; (2) the optical background is already quite low at nanosecond intervals, with typically less than one background photoelectron per interval; and (3) multiple paths through the atmosphere will spread pulses out and make it more difficult for the system to function with proper accuracy.

Another reason to prefer optical methods over radio SETI is that it is much easier to produce a narrow beam. In crossing interstellar space, a signal will travel many trillions of kilometers. If the sender were to broadcast in all directions simultaneously, i.e. omnidirectionally, then the power required would be prohibitive at any wavelength. George Swenson of the University of Illinois at

Urbana–Champaign has calculated that if a radio transmitter were 100 light-years away and radiated its energy in this manner, it would require 5,800 trillion watts to provide us with a detectable signal; an amount which, Swensen points out, is more than 7,000 times the total electricity-generating capacity of the USA. And (as discussed in earlier chapters) there is little chance of there being a communicating civilization as close as 100 light-years. Thus, directionality of the beam is essential. In general, the narrower the beam, the better, but in targeting a particular star system it should be confined in order to deliver all of the energy inside the radius of the star's habitable zone. A beam *that* narrow can only be achieved at short wavelengths below, at, or near the visible regime. As noted earlier, the attainable beamwidth is approximately the wavelength that is being transmitted divided by the diameter of the transmitter's antenna. The wavelength of light is six orders of magnitude shorter than it is for microwaves. So in a targeted approach the physics of beamwidth supports the use of lasers rather than radio frequencies.

Although Charles Townes, who co-invented the laser, and Robert Schwartz first suggested the idea of searching for optical signals from extraterrestrials in a paper in the journal *Nature* in 1961, the fact that laser technology was not nearly as mature as radio-frequency technology meant it was several decades before optical SETI caught on. Now microwave detectors have almost reached their fundamental limits in terms of noise, but there is significant scope for improving lasers. Another reason for the delay in pursuing optical SETI was the insistence on comparing radio frequencies with continuous-wave lasers, a preference for small laser and receiver apertures, and a study which showed radio frequencies to be more appropriate. This Cyclops report is available from the SETI League (www.setileague.org). Figure 12.1 shows a table from the report which well

	OPTICAL		INFRARED		MICROWAVE	
PARAMETER	A	B	A	B	A	B
Wavelength	1.06 μm	1.06 μm	10.6 μm	10.6 μm	3 cm	3 cm
TRANSMITTER						
Antenna Diameter	***22.5 cm***	***22.5 cm***	2.25 m	2.25 m	100 m	3 km
No. of Elements	1	1	1	1	1	900
Element Diameter	22.5 cm	22.5 cm	2.25 m	2.25 m	100 m	100 m
Antenna Gain	4.4×10^{11}	4.4×10^{11}	4.4×10^{11}	4.4×10^{11}	1.1×10^{8}	9.8×10^{10}

Figure 12.1. The Cyclops Report – Transmitter side. The data is taken from Table 5-3, page 50, July 1973 revised edition (CR 114445) of the Project Cyclops design study of a system for detecting extraterrestrial life. This study was prepared under the Stanford/NASA/Ames Research Center 1971 summer facility fellowship program in engineering systems design.

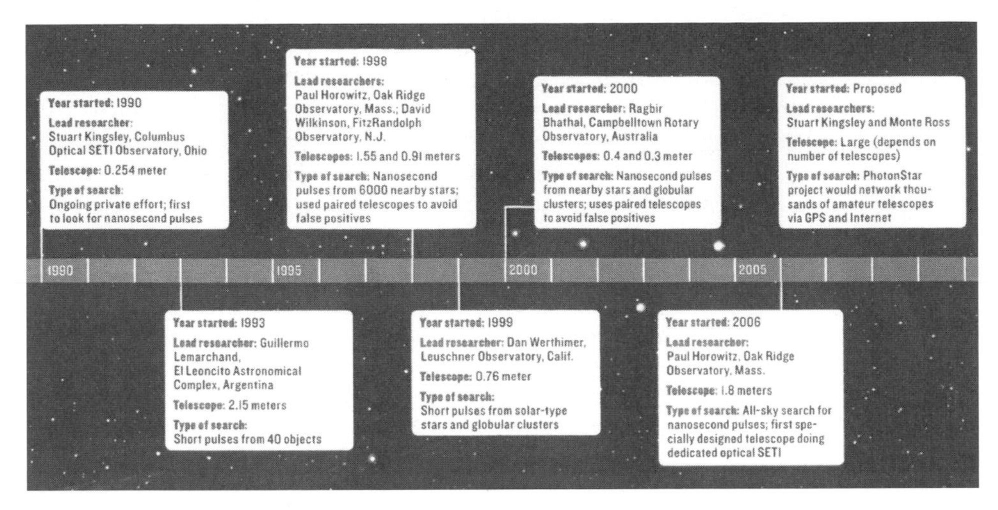

Figure 12.2. Optical SETI through the years.

illustrates the bias. The maximum collector diameter for a laser signal was only 22.5 centimeters, whereas the antennas for radio frequencies were up to 3 km in diameter. But Cyclops did not consider short-pulse lasers, even although there was literature available - including a paper by myself published in a leading journal in 1965.

SETI investigators began to take lasers seriously only after decades of listening at radio frequencies failed to produce a result. Figure 12.2 shows a timeline of efforts in optical SETI experiments. An initial experiment was undertaken in 1990 by Stuart Kingsley who recognized, in discussion with myself, that no one was looking in the optical spectrum and, in particular, no one was looking for nanosecond-type pulses. However, his 0.254-meter-diameter telescope and associated equipment constrained his measurement capability. In 1993 Guillermo Lemarchand in Argentina looked for pulses using a 2.15-meter telescope. Being in the southern hemisphere, he examined different stars than those available to observers in the USA. His observing time was limited, but represented a significant step towards placing optical SETI on a par with radio frequencies in terms of seriousness of intent.

In 1998 SETI investigators with resources began to look for nanosecond pulses. Paul Horowitz of Harvard built a nanosecond detection system that used a telescope at the nearby Oak Ridge Observatory in Massachusetts. He and his team operated this system intermittently for several years. Later, David Wilkinson at the Princeton Fitz-Randolph Observatory studied the same star at the same time in order to aid in distinguishing a real signal from a false alarm. The Harvard team used a 1.5-meter reflector, and about one-third of the light was deflected into the SETI receiver. The Princeton team's telescope was only 0.9 meters in diameter but they used almost all of the collected light. The two installations therefore had a similar amount of light available. Harvard passed the incoming light through a beamsplitter onto two Hybrid Avalanche Photodiodes (APDs) whose outputs fed high-level discriminators which had levels corresponding to roughly 3, 6, 12 and 24 photoelectrons. Approximate waveforms could be recorded to a precision of 0.6 nanoseconds by time-stamping level crossings. Coincident pulses triggered the microcontroller to record the arrival time and waveform profiles. A 'hot event veto' eliminated a class of high-amplitude bipolarity signals apparently due to breakdown effects in the APDs. Pulse counters and miscellaneous electronics allowed test apparatus to confirm proper operation. In particular, fiber LEDs were used to test the coincidence electronics prior to a search. The receivers were set to a sensitivity of 3 photoelectrons in each 5-nanosecond window. Taking into account the optical system losses and quantum efficiency, the receiver sensitivity signal level for the Harvard system was estimated at 100 photons per square meter and for the Princeton system at about 80 photons per square meter. These numbers apply to the visible spectrum between the 450 and 650 nanometers, for which the quantum efficiency is about 20 per cent; outside this range it is closer to 1 per cent. During 2,378 hours between October 1998 to November 2003, a total of 15,897 observations were made of 6,176 stars, typically for periods of between 2 and 40 minutes per session. The targets included all the main sequence dwarf stars between spectral classes A and early M within a radius of 100 parsecs and, owing to physical limitations, at celestial declinations between -20 and +60 degrees. A number of false detections occurred, including with the domes closed. In some cases these were equipment problems, but most appeared to have natural explanations, including the possibility of strikes by cosmic-ray muons. A lot was learned about the practicalities of doing optical SETI. In particular, a second observatory coupled with precise event timing completely eliminates background events and provides for the strongest possible confirmation of a signal. The lessons learned were applied in designing the dedicated all-sky system discussed later in this chapter.

A large number of small telescopes scattered geographically, as in the PhotonStar concept, will have no background problem. The Harvard and Princeton experiments showed that making the instrument robust against many failure modes with reliable software, good telemetry, and optical fibers for lightning protection is key for nearly automatic operation with low maintenance. The need for diagnostic checks prior to each observation is self-evident. In an experiment seeking a rare result, it is essential that the apparatus operate at its best during an observation. End-to-end testing with a source in the far field is highly desired.

In 1999 Dan Wertheimer of the University of California at Berkeley began to look for optical signals using the 0.76-meter telescope of the Leuschner Observatory in California. In 2000 Ragbir Bhathal in Australia resumed the investigation of the southern hemisphere. He designed a system to detect short pulses and used a pair of telescopes to avoid false alarms. His telescopes were 0.3 and 0.4 meters in diameter, and were located at the Campbelltown Rotary Observatory.

One can see from this summary that optical SETI has attracted far less

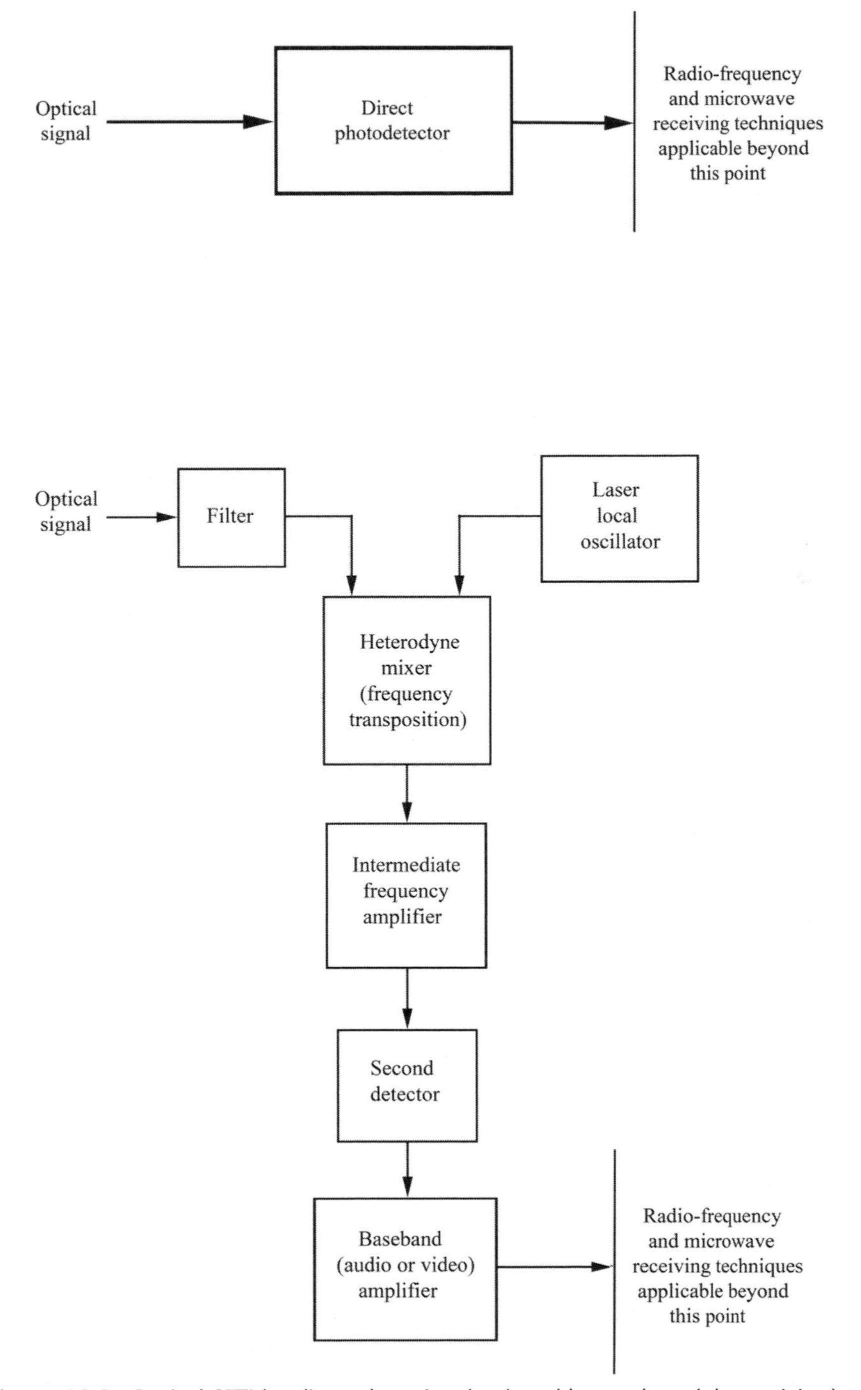

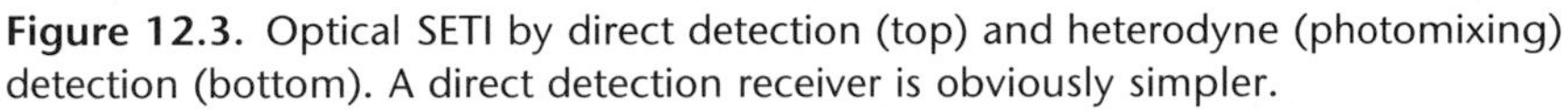
Figure 12.3. Optical SETI by direct detection (top) and heterodyne (photomixing) detection (bottom). A direct detection receiver is obviously simpler.

resources than were made available to radio-frequency research. The dedicated all-sky system developed by Harvard represents a leap forward in experimental investigations, but its collector area is far less than could be achieved relatively cheaply by the 'photon bucket' approach. Figure 12.3 is a general block diagram of an optical receiver. The two basic methods are either direct detection or heterodyning (photomixing). As is evident from the figure, heterodyning is far more complex. As noted earlier, direct detection is the technique of choice for optical SETI not only because it is simpler and avoids many problems, but also because it can come close to the performance of an optimal heterodyne system. To summarize: direct detection eliminates the need to know the exact wavelength, does not require all parts of the received signal to be in phase, and is insensitive to Doppler shifts. The low background due to examining time slots of nanosecond duration allows the use of large collectors. The efficiency is improved by eliminating narrowband optical filters. All that is required of the optics is that the field of view be narrow enough to exclude other stars as bright as the target star, and that the light be focused on a spot no larger than the input of the photodetector. The fact that the phase information is not needed makes it possible to use less accurately figured optics that are considerably cheaper than optics of comparable size intended for imaging. In particular, large-segment low-cost mirrors can be used without the need for adaptive real-time alignment correction to eliminate the fluctuations caused by the light's passage through the Earth's atmosphere.

By way of example, Figure 12.4 shows a 4.2 meter-diameter optical system capable of being steered at speeds sufficient to take only a few seconds to slew from one star to the next. The main mirror is made up of 18 hexagonal segments, each approximately 0.8 meters across. The design is based on three assumptions. The first assumption is that the sender will aim its signals only at star systems that are likely candidates for hosting intelligence (with us included on their list). The second assumption is that a signal will be sent in such a way as to prevent accidental discovery by others. Only when one star system at a time is encompassed by the beam can accidental discovery be minimized. The broad beams typical of radio-frequency signals would inevitably cover many star systems simultaneously. Although sub-microradian radio-frequency beams could be achieved using enormous antennas, why would the sender proceed that way when a small-diameter optical system could readily achieve the same purpose. The strength of the laser can be tailored so that when the signal reaches the target it will be at the level consistent with the expected capability of a society possessing the technology to make a detection. The third assumption is that there is no need to transmit on a continuous basis to a single star system, so the signal will be offered to the receiver only for part of the time. The sender will slew the laser sequentially from one target star to the next, covering all of the likely candidates in a tiny fraction of the time it takes for the signal to travel through space. This enables a single laser to cover many hundreds of candidates. (There is no need to build a separate laser for each one.) By applying these assumptions, we arrive at an optical receiver design that satisfies the requirements within our

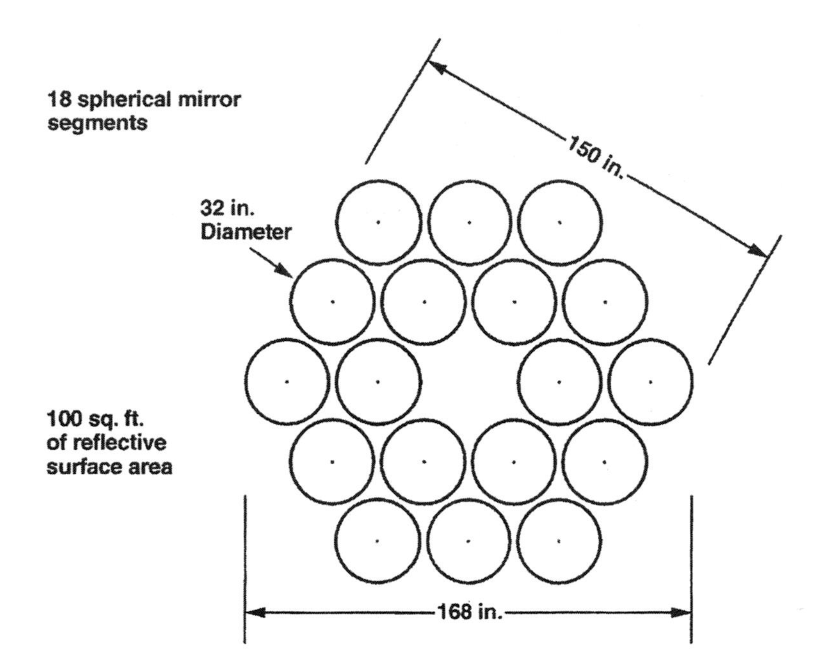

Figure 12.4. A segmented primary mirror for an optical telescope.

technological capability, but has not yet been implemented. It has four major features:

- The receiving apparatus must be of at least a certain size, since otherwise the signal will go undetected.
- The receiving apparatus can rapidly slew from one star to another to inspect all likely sources in a time period during which a sender, when aiming a laser at us would (if existing) send a signal.
- A short-pulse detection system enables the signal to outshine the host star.
- It must be dedicated to this task so signals (if present) are not missed.

No existing laser receiver satisfies these requirements. The closest match, the new Harvard system, meets two requirements but falls short in terms of receiver size and the ability to slew rapidly.

Non-imaging telescopes that have large collectors include the old Mount Hopkins 10-meter-diameter instrument in Arizona originally utilized for Cerenkov radiation studies, and the McDonald Observatory in Texas which has an 11-meter-diameter collector that consists of 91 segmented mirrors. The segmented primary mirror in Figure 12.4 provides a collection area of 9.2 square meters. As the targeted stars will be only a few degrees apart, rapid slewing is possible by electro-mechanical means. Our technology is already sufficient to build such a collector. A 10.4-meter-diameter mirror consisting of 36 hexagonal segments, each of which has its own computer-controlled actuators for maintaining the mirror's shape as it turns to track stars, was inaugurated in 2007 on the island of La Palma in the Canaries. To facilitate imaging, this aptly named Great Canary Telescope provides much higher resolution than is necessary for SETI. It should retain the record for size until the European Extremely Large Telescope enters service in 2017 with a mirror 42 meters in diameter made up of 984 hexagonal segments. Perhaps we will require optics of such size to detect an alien laser, but in the meantime we will have to make the best of the new Harvard system.

All-sky optical SETI at Harvard

Optical SETI entered a more mature era with the introduction in 2006 of a dedicated facility with which to detect nanosecond pulses. This telescope and signal detection and processing capability were designed by a team at Harvard led by Paul Horowitz and built at the Oak Ridge Observatory in Massachusetts. Significantly, this system is designed to search by scanning the entire sky rather than targeting individual stars. It exploits advancements in optical and laser technologies over the past two decades. For example, it uses improved quantum efficiency photodetectors with nanosecond response times. A key feature of the system is the use of multi-pixel photomultiplier tubes to examine more of the sky simultaneously. The primary mirror is 1.8 meters in diameter and the secondary mirror is 0.9 meters in diameter (Figure 12.5). The non-imaging optics (that is, a 'photon bucket') were cheaper than a comparable imaging system. The telescope is designed to cover the northern sky in 150 days. However, this assumes that a laser is pointing in our direction continuously in a pulse mode. If, as is more likely, the laser points at us only intermittently as it works through a list of possibly thousands of stars as potential hosts of civilizations, then the probability that it will be sending to the Sun at the same time as we look at the host is miniscule. Nevertheless, the increase in capability from sporadically examining a small number of stars to conducting an all-sky survey is a significant step towards a serious search for alien signals. (In Chapter 16 we will discuss a strategy which overcomes the issue of reliably detecting a signal when the transmitter and receiver are not always pointing at each other).

The Harvard system divides the sky into patches of 0.2 × 1.6 degrees. It observes each for 48 seconds before moving to the next. The telescope is able to move only in declination, and as the Earth's rotation sweeps the field of view in right ascension the system scans a strip of declination on a continuous basis. As discussed earlier, photomultipliers convert photons into electrons with very little added noise. A series of stages amplify the number electrons in a cascade until they exceed the electrical noise and are output. A multi-pixel photomultiplier divides the collection area into tiny squares, with each channel acting like a separate detector. In this case there are 64 squares per tube. This facilitates looking at more than one star system at a time. Because the electronics are looking independently at each nanosecond, an enormous amount of processing

Figure 12.5. Harvard's all-sky search telescope for optical SETI.

takes place during those 48 seconds. All the signals from each pixel are fed into 32 microprocessors custom-designed for the project by Horowitz's graduate student Andrew Howard. These chips crank through the data at a rate of 3.5 trillion bits per second in search of a large spike in the photon count that could be a laser pulse from space. The receiver's detectors are divided into two arrays so that if one array detects an interesting signal it can be checked against the other one. With this implementation, internal system noise is unlikely to cause a problem. If a pulse is detected in one channel, it will have to be detected in the same nanosecond in the second channel. Stray noise from outside the system

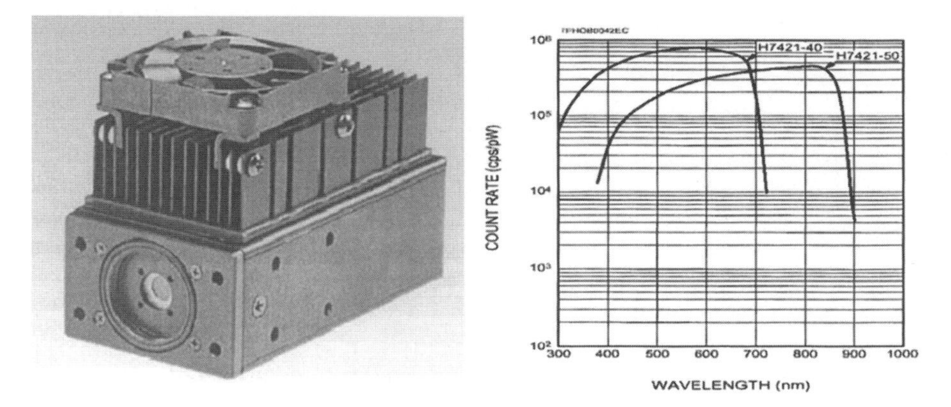

Figure 12.6. Response versus wavelength for a Hamamatsu H7421 photomultiplier photon counter.

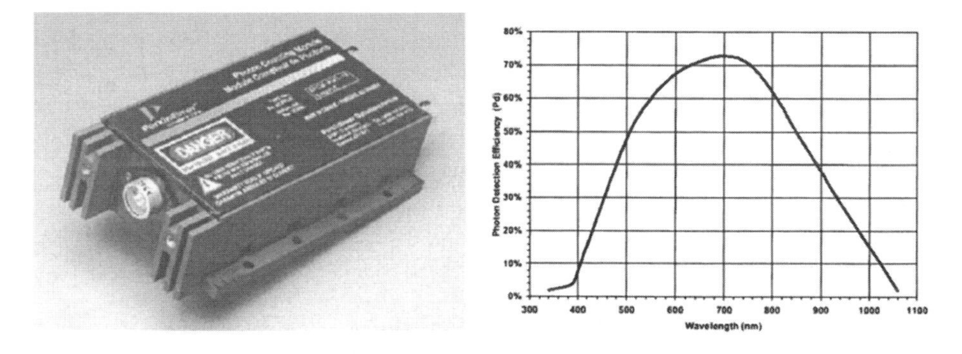

Figure 12.7. Response versus wavelength for a Perkin-Elmer solid-state avalanche photodiode single-photon counter module.

could still cause a problem, but adding a second system some distance away which simultaneously examined the same piece of sky would enable random false detection events of this type to be discarded. Whilst two photodetector channels reduce false detections to perhaps one occurrence per night, adding a third channel reduces the rate to only one per year or so.

With the recent introduction of Avalanche Photodiodes (APDs) as alternatives to photomultipliers, APD arrays may substitute for photomultipliers, especially in the near-infrared in which the quantum efficiency of photomultipliers drops off rapidly whereas APDs still offer high quantum efficiency and gain with little added noise. Figures 12-6 and 12-7 show the responses of sensitive photomultipliers and APDs with wavelength. Photomultipliers have been made with 40 per cent quantum efficiency at green, but typically are less than 1 per cent at wavelengths of 1 micrometer in the near-infrared. This low quantum efficiency makes APDs attractive even if they add a small amount of noise by having a quantum efficiency of up to 80 per cent. Future improvements in solid-state detector arrays are expected to facilitate single-photon detection using an APD in the Geiger mode, where it acts as a trigger. In this mode, single-photon detection events can improve sensitivity. Although it will also render the detector essentially inoperative for several nanoseconds, this should not pose a serious handicap to a system with a low duty cycle. An example would be a system of 1,000 slots, with each slot of 1 nanosecond contributing to a total duration of one microsecond. Reference pulses would be provided every 3 microseconds. The first microsecond after a reference pulse would not be used, nor the microsecond before the reference pulse. Since the data pulse can only happen in the middle microsecond, this prevents any problem arising from the Geiger mode operation. The Geiger mode detector recovers in much less than a microsecond when it is triggered by a nanosecond pulse.

Beamwidth and habitable zones

One of the benefits of a laser is that it can form a very narrow beam. This facilitates sending a beam just wide enough to cover the habitable zone of the target star, making the most efficient use of the energy. When considering using narrow beams to signal a star system, there are several issues which affect the pointing accuracy: namely pointing misalignment, pointing jitter, and point-ahead error. The first two represent minor imperfections in hardware and limit how narrow the beamwidth can usefully be (Figure 12.8). It is desirable that the combined effect of misalignment and jitter produce a pointing error no greater than a small fraction of the beamwidth, say no more than 10 per cent. As the beamwidth gets narrower, this becomes harder to accomplish. The third issue is an error arising from imprecise knowledge, and could well be the largest uncertainty. It results from the fact that the targeted star is not only so distant that a signal traveling at the speed of light may require centuries to reach it, the star is also moving independently through space. If the laser were to be aimed precisely at the star, the star will have moved by the time the beam arrives. Both the speed and direction of a star must be known very accurately to calculate how much to offset the laser. Obviously, this depends on the relative positions of the two stars in the galaxy. The worst case would be where the point-ahead error was greater than the beamwidth, and the signal missed the target completely. Since we are discussing such small angles and rapidly moving targets at such great distances, serious errors can result. If each of the three factors were limited to 10 per cent uncertainty of the beamwidth, and all errors tended in the same direction then a 30 per cent increase in the beamwidth would be needed. This would translate (by squaring 1.3 to get 1.69) to a 69 per cent increase in laser power in order to maintain the number of photons per square meter. A plausible calculation of the point-ahead accuracy indicates that it must be 1 part in 100,000 to avoid an impact on the minimum useful beamwidth. We have considered links in which the minimum is as small as 10^{-7} radians. At 10 per cent, we need to know the point-ahead to less than 10^{-8} radians. At a distance of

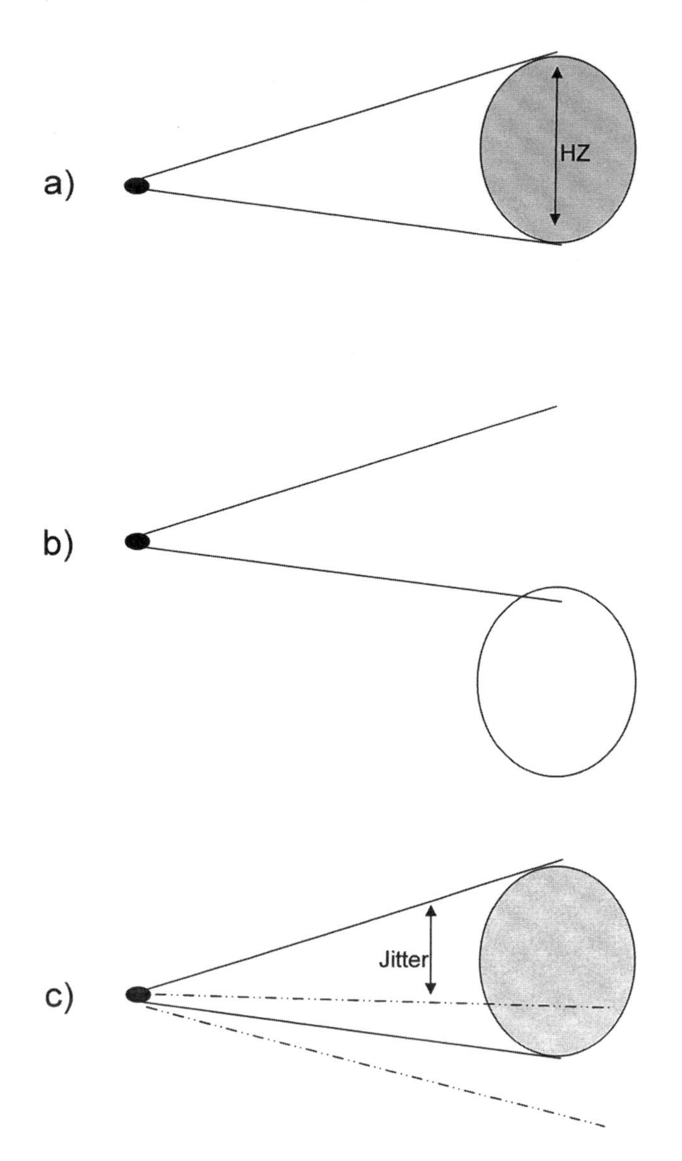

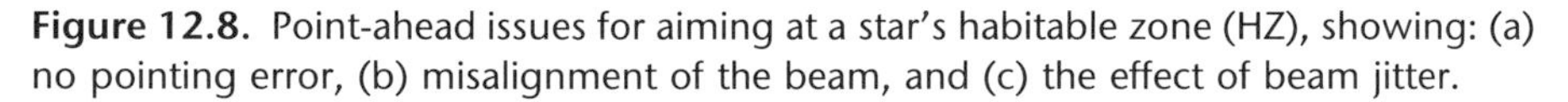

Figure 12.8. Point-ahead issues for aiming at a star's habitable zone (HZ), showing: (a) no pointing error, (b) misalignment of the beam, and (c) the effect of beam jitter.

100 light-years, this is a displacement perpendicular to the line of sight of 4.5×10^{-4} light-years, which is large relative to the diameter of a habitable zone of a solar-type star. An accuracy of 1 per cent is preferable. Whilst this may be difficult to achieve, it should be feasible for an advanced civilization that will know much more about its stellar neighborhood than we do at the moment. The point-ahead angle will be different for each targeted star. However, this is a transmitter-only issue. In seeking a laser signal we look with a field of view of

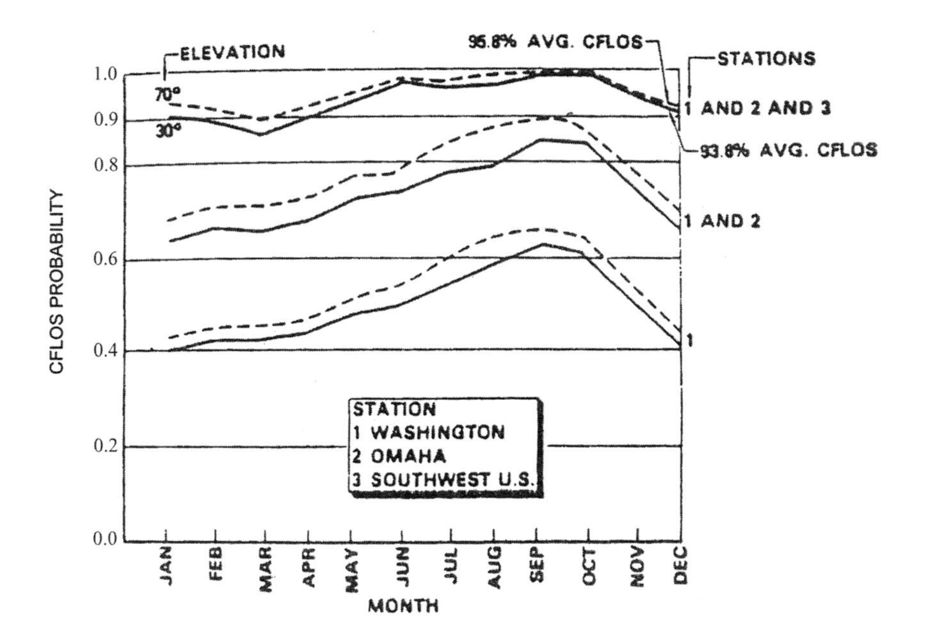

Figure 12.9. The probability of a cloud-free line-of-sight for three specific locations in the United States. The individual locations have probabilities of only 0.4 to 0.8, but there is an average probability of 0.938 of one of the three being clear.

typically 1 milliradian, which renders insignificant a shift measured in microradians, and in any case we see the light as it *arrives* – we do not need to make any allowance for how its point of origin has moved since the light began its journey; at least not until we attempt to reply, at which time we must make the corresponding calculation.

Weather and other issues

Although there would appear to be no celestial events which could be mistaken for a 1-nanosecond laser pulse, terrestrial weather poses a problem. Obviously the best site to place an optical receiver would be a location which has clear skies, but the situation improves if a number of telescopes are distributed geographically and work in a coordinated way. Figure 12.9 shows the percentage of time for cloud-free line-of-sight on the basis of three widely separated sites in the USA, with a figure of almost 95 per cent being achievable.

At optical wavelengths there is a certain amount of loss arising from light being scattered and absorbed. Optical scattering is caused by interstellar dust grains. This is directionally dependent and can be significant over great distances, but is minor within the search radius considered for SETI. The effect on a laser pulse is to reduce the pulse height and simultaneously yield delayed tails on two timescales: a close-in tail from forward scattering by large dust grains, and a much longer tail from diffuse scattering. The major leading edge retains its pulse shape, but having lost photons to scattering its amplitude is reduced. For signals sources 1,000 light-years away, the maximum expected reduction is 40 per cent. At slightly longer wavelengths than optical, such as near-infrared, this effect is reduced. Optical SETI by short-pulsed laser should therefore be perfectly viable within the radius we have selected for our search.

Part 4: Possibilities in SETI

13 The PhotonStar project

The PhotonStar SETI project to detect alien laser signals involves a large number of small telescopes acting together in a geographically dispersed array. Figure 13.1 shows a conceptual view of the system. In acting together, looking at the same star system at the same time, the large overall collection area increases the chance of detecting a signal, should one be present. The position of each telescope can be located by GPS so that the differential distance from a given target star to each telescope can be determined to calculate timings, and the Internet used not only to coordinate the switching of one target to the next but also to send the data to a central station. This became feasible with the advent of relatively low-cost single-photon detector technology for use by amateurs. The timing of the receivers can yield better than 10 nanoseconds accuracy – i.e. once

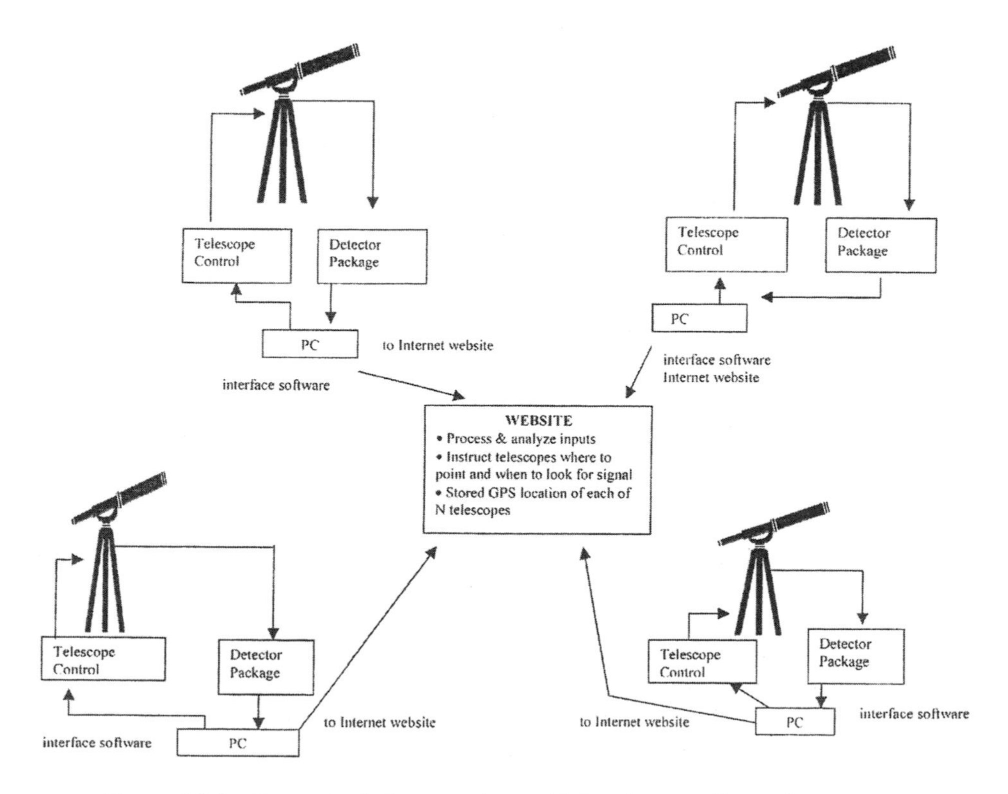

Figure 13.1. Conceptual diagram of a multiple-telescope PhotonStar system.

the physical locations of the telescopes had been allowed for, if a pulse were to occur then every receiver that detected it would measure the time of detection as being within 10 nanoseconds of any other telescope in the array, which is sufficient. There are thousands of amateurs with telescopes of 20 cm or more in aperture. Arraying offers a total collection area greater than the largest optical telescope. The signal photon flux from the target is obviously the same irrespective of whether there is a one collector with an area of 10,000 square meters or an array of 10,000 collectors each providing an area of 1 square meter. If signal photons are received by the single large collector, scaling dictates that the same number of photons will be collected in total by the array. If the flux is low, then a few detectors of the array will still receive photons even if the majority receive nothing. Poisson statistics will apply, and some detectors must receive a photon such that the average number of photons being detected will be the same whichever architecture we use. We have no control over the incoming photon density. In this way, each detector sends its data to the central station via the Internet for analysis in real-time. Specific directions prior to observing will enable each telescope to look at the proper star and know the times at which to do so. Each receiver has its own set of coincidence detectors to reject internal noise results. Although it is unlikely that a pulse detected simultaneously by widely separated receivers is not extraterrestrial, it will be thoroughly examined to determine whether it could have been otherwise.

Using plausible numbers, if the sender is able to confine the beam to the habitable zone there will be roughly 0.01 photons per square meter. This photon flux density will ensure that at least one photon per pulse reaches the receiver. It is based on the sender expecting that the receiver collector will have an area of at least 100 square meters. But optical system losses and the quantum efficiency of the detector will require at least five photons per pulse to reach the receiver. Hence laser energies of the order of 1,000 joules per pulse are needed (without considering optical path losses that could contribute a factor of two). For a rate of 10 pulses per second, the sender will need a laser with an average power of 20 kilowatts. This is based on the beam being confined to precisely match the habitable zone. The beam should not be made narrower than this, or it might miss the receiver. If it is wider, then either the power must be increased to compensate for the dilution of the energy or the recipient will make the receiver larger than the minimum of 100 square meters in recognition of the difficulty a sender may face in precisely pointing a beam which is tailored to the habitable zone of a solar-type star. What we can certainly conclude, is that we cannot reasonably expect a few meters of collection area to be sufficient. A telescope with a diameter of 30 cm has an area of about 0.07 square meters, so an array of 1,000 such telescopes would provide an area of 70 square meters and 10,000 telescopes would provide 700 square meters. Given the impracticality of an enormous single collector, arraying would seem to be the way to go.

The requirement for large collection areas raises two points. First, the energy-per-pulse requirements for a useful pulse rate are readily attainable and do not impose a serious burden on the transmitting society. They could have a number of systems, each of which is signaling to hundreds of stars in sequence (Chapter 16). Second, a potential recipient must make a serious effort to detect such a signal, but this, too, is readily attainable.

Minimum useful beamwidth

The minimum useful beamwidth is that required to precisely match the habitable zone of the target star. It will be different for stars of various spectral classes, and at differing distances from the sender. In the case of the Sun, the habitable zone has a diameter equivalent to approximately 30 minutes of light travel. If, as we expect, the sender requires us to make a serious effort, then the photon density will be such that the number of photons per pulse will be less than the number of square meters in the zone, and the photons collected will certainly be less than the number of receivers in an array. Whereas a single large collector may receive ten or so photons, in the array of small collectors with the same total area typically ten of the receivers will report single-photon detections to yield the same photon count. However, in the case of the array, each transmitted pulse will be detected by a different set of ten receivers. But the single large collector sees more background light than any of the collectors of an array by a factor which is the ratio of its collection area to that of any given smaller one. The background will therefore be less of an issue for smaller collection areas. A receiver which does not presume any knowledge of the laser's wavelength becomes more feasible.

Participation of many

PhotonStar enables the participation of both amateur and professional astronomers at the cost of a laser receiver per telescope. If the receiver were to be standardized and thousands made, then the cost would be relatively low. The requisite software can be downloaded from the Internet. The data output has to be packaged and sent via the Internet to a central station that looks at the data from the full array in real-time. The system offers the attractions of: (1) avoiding the necessity of building a large-optics multi-million dollar system to seek extraterrestrial pulsed lasers; (2) it enlists anyone who wishes to participate; (3) due to its geographical diversity it is less susceptible to weather constraints; and (4) it can grow as more receiving stations are added. But it was impractical prior to the advent of GPS, the Internet and single-photon detector technology. The laser receivers should be designed to be readily used in conjunction with existing computerized telescopes, such as those by Meade and Celestron. If the software is user-friendly, then no special knowledge will be needed to participate. It ought to be possible to enlist thousands of these individual systems and operate them in a coordinated manner. The central station should be designed to enable the array to expand without disturbing the existing users. By knowing each small telescope's location precisely in terms of latitude,

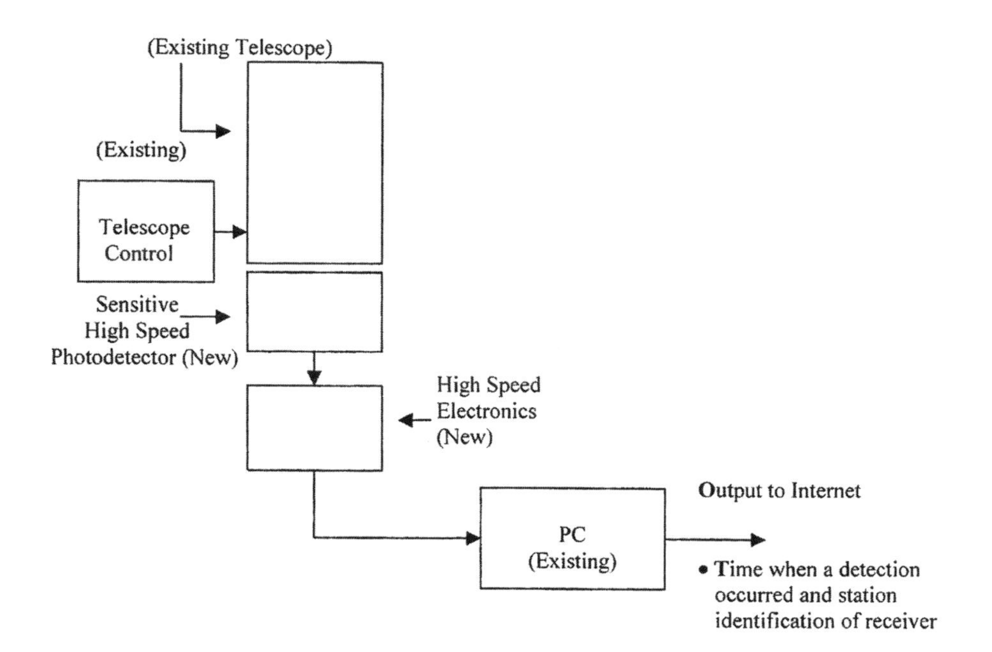

Figure 13.2. Receiver station hardware configuration of the PhotonStar system.

longitude and altitude, the central station will direct each telescope to enable them all to aim at the target star at the same time and work through its list of candidates. If a pulse detection occurs, the time of detection is sent to the central station, which notes how many receivers detected it at the same time. For the expected weak signal flux of optical SETI, a small number of receivers in the array will detect pulses and most will not. However, the average total number of photons detected should be the same as for a single large collector with the same area as the array. Each time a weak pulse is detected, a different set of receivers will detect the pulse. Because long-term monitoring of a fixed site by GPS can determine its position to within a few centimeters, the difference in distance for each telescope to a location in space can be calculated. The range will vary over time, as the Earth's rotation causes the angle of viewing to change, so all of this must be tracked in real-time. Figure 13.2 shows the hardware configuration of a site, involving the telescope, the PC, and the single-photon detector. Each receiver has its own identification number to enable the central processor to determine which telescopes are reporting detections. The detectors are fast enough to distinguish 1-nanosecond pulses. At light-speed, 1 nanosecond is equivalent to a distance of 0.3 meters. If one receiver is 1,500 meters further from the source than another, it will detect a pulse 5,000 nanoseconds later. Once the geometries of the sites are normalized, the signal reception timings can be correlated to determine whether they were simultaneous. False pulse detection will occur with an array of small collectors, but the central processor should be able to sort out false signals through time of detection and lack of simultaneity.

Large Telescopes	Area (meters^2)
3 meter diameter	~7.0
10 meter diameter	~77
30 meter diameter	~700
Small Telescopes	
12 inch diameter	~0.07

Figure 13.3. Photon collection area.

N	Area (meters^2)
100	~7
1,000	~77
10,000	~700

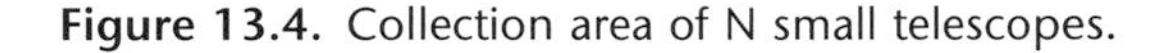

Figure 13.4. Collection area of N small telescopes.

Small telescope arrays versus one large collector

Figure 13.3 shows the collection area of a large-diameter telescope. Figure 13.4 shows a number of much smaller telescopes attaining an equivalent area. Arraying offers a number of benefits over a single large telescope. Obtaining time on large telescopes for SETI is always difficult. Whilst a dedicated system can be built, as Harvard has done, it involves significant initial cost and is subject to outages due to weather and essential maintenance. A diverse array of small telescopes makes use of an existing infrastructure of telescopes and PCs, is able to expand as stations are added without the need for reconfiguration, and is less susceptible to weather. With a large number of collectors tied to a worldwide network, new strategies can be used. In particular, instead of having the entire array work through the target list in concert, it would be possible to assign subgroups to particular stars and upon a detection being made the entire array swings onto that star to maximize the effort. The receiving stations can operate automatically, with the central station controlling where the telescope points and being alerted by any detections, so once set up they do not require the constant attention of the owner.

As the technology improves, the best choice of detector today may not be the best tomorrow, so an array might incorporate a variety of detectors. Today's commercial single-photon detectors are made by Hammatsu and Perkin-Elmer and sold as Single Photon Detector packages. Candidates include the HPMT (Hybrid Photoelectron Multiplier Tube) and the solid-state Hybrid-APD. High-speed circuitry is needed in order to properly process the pulse output to precisely determine the time of signal detection. If the pulse is broadened owing to insufficient detector/electronics bandwidth, the accuracy of the time of detection will be impaired. This, in turn, will make it harder to avoid false alarms. The likelihood of false detections increases as the pulse width broadens. For example, if a receiver seeks 1-nanosecond pulses and the receiving capability can only resolve 10 nanoseconds, it is uncertain which of the nanosecond periods it actually occurred in. Any background photons that were in the 10 nanosecond period could falsely be counted as signal photons, as on the basis of time received the receiver was incapable of telling the difference. A large collection area enables a relatively low energy per pulse laser to be detected at a range of many light-years. Figure 13.5 shows a laser of only 10,000 joules per pulse being detected at 200 light-years. For 10 pulses per second, the laser would need an average power of 100,000 watts. Signal detection is determined by a sufficient number of the stations detecting a photon at the same time. Let us say it is calculated from the spectrum of a star that in each 0.1 second interval this background should prompt two of 100 stations to detect a false pulse at the same time. Suddenly ten stations report a pulse, making this an event worthy of examination. If it is a true signal, it can be expected that a different set of about ten stations will report when the next pulse arrives. Other factors being equal, the stations that report ought to be random. If a given station is reporting detections in excess of the average then it is probably sending false alarms owing to excessive internal noise or is malfunctioning. The signal processing at the central station must examine every nanosecond period and determine whether,

Nanosecond pulse

Area of beam at 200 lys ~7 x 10^{23} meters squared

10 meter diameter collector for ~70 meters squared for receiver

Quantum efficiency at 0.5 micrometers (green) ~0.2

Assume 5 photoelectrons needed for detection.

E_r = hfn ~4 x 10^{-19} joules per photon x 25 photons received per pulse = 10^{-17} joules per pulse

E_t = transmitter energy per pulse = E_r x A_t / A_r where A_t / A_r is 10^{21}

Resulting in 10,000 joules per pulse (=100,000,000,000,000 watts for 1 nanosecond)

Figure 13.5. Laser energy per pulse requirements for a link at 200 light-years.

based on the data supplied by many receivers, there were real pulses. Fortunately, only the central station requires such computing power, the individual stations do very little processing and this makes the system feasible.

Although there is currently no effort underway to implement the PhotonStar array, it is likely that this approach will eventually be attempted owing to the fact that it is a viable way of achieving a large collection area for short-pulse signals. And, as SETI@home proved, the public is willing to participate directly in the search for extraterrestrial signals.

14 Key issues for SETI

Several key issues for SETI must be addressed before we can conduct an efficient search. These include the number of potential star systems as a function of distance, the sender's choice of broadcast versus targeted signaling, the possibility of sensor probes, and whether we ourselves should be sending signals – the significance of the latter issue being that an alien civilization might be reluctant to reveal its existence to us.

It makes sense for a sending civilization to confine the beam of its signal to just the habitable zone of the targeted star. This minimizes the power requirement of the transmitter. Only if the target is so far away that the beamwidth needed becomes too narrow to be reliably held on the habitable zone, must the power be raised above the theoretical minimum. Another factor is the choice by the sender of photon density at the receiver, which is the number of photons per square meter. As discussed earlier, the habitable zone of a solar-type star is less than one-thousandth of a light-year in diameter. Given a transmitting antenna that is 10 meters in diameter, we can readily calculate the beamwidths attainable for different wavelengths. For a wavelength of 0.5 micrometer in the visible region of the electromagnetic spectrum the beam will be about 5×10^{-8} radians, and for a wavelength of 2 micrometers it will be 2×10^{-7} radians. Simple geometry shows that a beam of this width would spread out to span one-thousandth of a light-year (and thereby precisely match the habitable zone) after distances of 20,000 light-years and 5,000 light-years, respectively. (Notice that for a 1-meter-diameter optics transmitting at 2.0 micrometers, dividing the wavelength by the diameter of the telescope we find that it could be matched to the habitable zone out to 500 light-years.) To calculate the energy per pulse, we require the area of the habitable zone, which is about 70×10^{22} square meters. For an energy density of one photon per square meter at the target, the transmission pulse would have to provide 70×10^{22} photons, which is 700 joules. Optical path losses could contribute a factor of two or so. Call it a maximum of 2,000 joules. This should be readily attainable by an advanced society. The minimum size of collector for one photon per square meter is 10 square meters. If, however, the energy density were one-tenth of this intensity then the collector would have ten times the area.

Hence, once we recognize that targeting the habitable zone of a star ties down the transmitter requirements, we can calculate the minimum size of collector required to achieve detection. This line of reasoning therefore removes some of the uncertainty of the 'link' parameters. The energy per pulse at transmission depends on the photon density the sender desires the pulse to possess at the

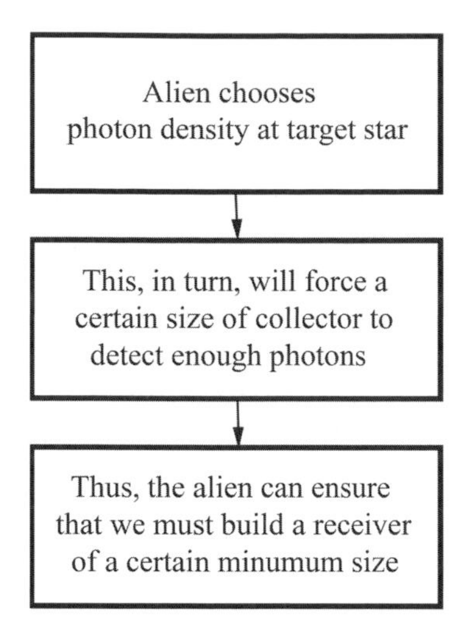

Figure 14.1. The alien civilization chooses the receiver size for us.

target, which in turn defines the minimum receiver size required for detection. Thus, if the aliens require us to build a receiver of a certain minimum size they can accomplish this by choosing the energy sent per pulse. They control the receiver size by controlling the photon density. They control the photon density at the target by confining the beam to a well-defined area, the habitable zone, whose angular measure can be calculated from the distance to the target star. Figure 14.1 shows this reasoning. The minimum useful beamwidth is that which precisely matches the habitable zone. As the beam is progressively narrowed in order to reach more distant targets, the area illuminated is a constant. If we ignore path losses due to scattering effects, which should be small, the laser power need not be increased until the pointing uncertainty becomes significant.

Let us digress to address a question that is often asked: Why can't the laser just provide a brief flash that could be seen by the naked eye? The simple answer is the difference in collector size. If it takes 2,000 joules per pulse for a 10-meter-diameter collector to intercept the signal, and to a first approximation the eye is 0.01 meters in diameter, the transmission power would have to be one million times greater. And since about 100 photons are required for the eye to detect an event, in comparison to 10 for a sensitive photodetector, this would add another order of magnitude. There are other factors too, but it is already evident that there is no hope of seeing an alien signal simply by peering up at the sky.

Another factor in choosing the size of an individual collector in an optical array is that it should allow less than one background photoelectron per measurement period, which we take to be 1 nanosecond, and avoid using a narrowband filter to screen out the background. Rather than impose a wavelength constraint, let the system rely on temporal screening. Since a pulse has to occur in a number of receivers at the same time, the receiver parameters can be chosen to provide a low likelihood of a false detection. The formula:

$$R = (r_1{}^2\tau)(r_2{}^2\ \tau\ )(r_3{}^2\ \tau\ )\ldots(r_n{}^2\tau)\tau$$

gives the rate of false coincidences. Using the background expected of a given star, and given its distance in light-years, we can calculate the rate of false coincidences. In fact, for a nanosecond interval the background will be very low. If the signal is approximately one signal photoelectron for each collector in the array, it will show a detection occurring in most of the collectors simultaneously as a true signal. Since when a signal is present one photon has to be detected, as long as the background is much less than one photon per measurement period this does not oblige the signal requirements to be any higher than for a situation in which there is no background at all. So if a society wishes a signal only to be seen by an antenna of huge proportions, they can make this a requirement by controlling the photon density. They may wish to expose themselves to discovery only by a technology that is sufficiently advanced to perform such a search. As we have noted, the necessary receiver size is dependent on the photon density, which is under the control of the sender. As larger collectors are more sensitive to background light, there is the temptation to use a narrowband filter to limit the background, but this also constrains a potential signal. An array of small collectors allows signal detection over a wide optical spectrum without letting the background starlight interfere with detecting a true signal. Note that by limiting the transmission energy the sending civilization is deliberately forcing the potential recipient to make a large receiver, which is better done by arraying than by building a single massive collector.

Some people have argued that any signal we receive will be what is called spread spectrum communications. This sophisticated technique applies to radio frequencies, and is used in cell phones. It requires a receiver capable of discriminating one carrier wave from another. It is less sensitive to interference, and because it sends pulses at different wavelengths it provides a more secure link. However, this technique is not feasible at optical and infrared wavelengths since the direct detector is wideband and does not distinguish individual frequencies.

Interstellar probes

Automated probes may be roaming the galaxy seeking electromagnetic radiation as a sign of intelligent life. If so, they would monitor the entire spectrum. In the 1920s Earth became a source of signals from AM radio stations. The energy, weak as it is, could be picked up by such a probe anywhere within a radius of 80 light-years. If it sent a reply, we might not receive this until late in this century. Or perhaps the probe simply gathers data without the authority to reveal itself. It could employ a dedicated laser link to relay anything it picks up to its home base, so remote that by the time the original signal had traveled that far it would be too weak to be detected. A probe would be an artificially intelligent entity, obeying general instructions but operating independently in real-time. It could

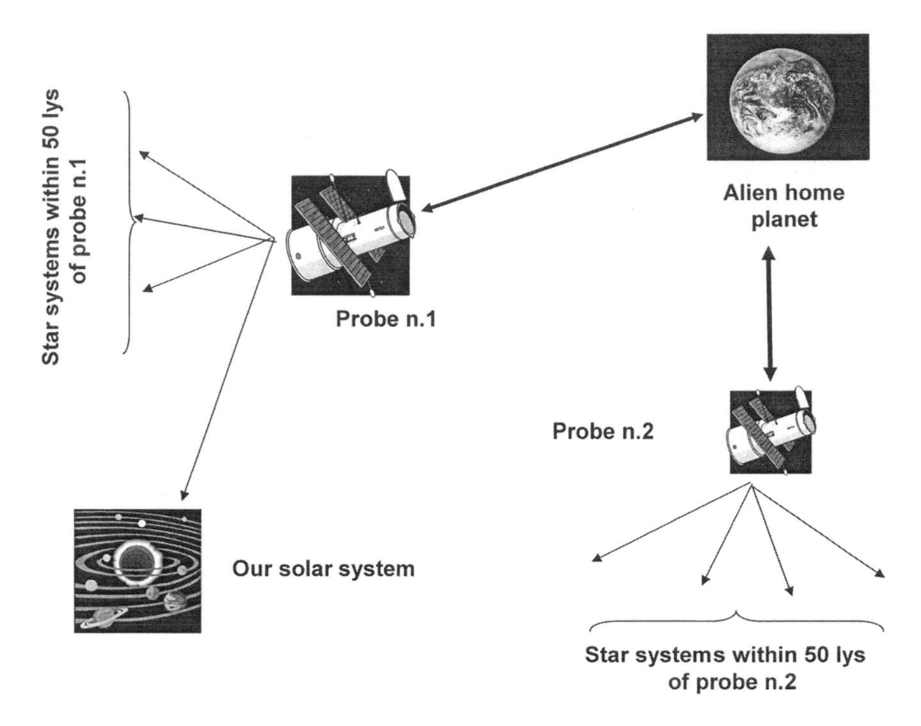

Figure 14.2. Interstellar probes may be used to search the galaxy.

be proactive, using a high-energy laser to target stars in its path which it judged to be potential hosts of intelligence. If a probes was able to travel at near-light-speed, then in the long term a flotilla could significantly extend the radius to which the parent civilization was able to explore. Alternatively, using a probe to issue signals would allow that civilization to protect itself. Figure 14.2 depicts this scenario. The difficulty facing us here, is that we are unlikely to detect a laser coming from seemingly empty space using a targeted search strategy - the only way would be to literally search the entire sky.

Broadcasting versus targeting

To design a system for receiving alien signals which follows a logical approach, we must first consider the question of whether that signal is more likely to be sent in a broadcast mode or will be aimed at one star system at a time. By broadcast, we mean using a transmitter that radiates its signal across a substantial fraction of the sky; in the worst case, the entire sky. A targeted approach will exploit knowledge of a wide range of scientific disciplines to select star systems having a reasonable likelihood of hosting intelligence. The mode deemed more likely has a major effect on the kind of receiving system we would use in the search. If the alien does not fear announcing its existence, and has no call for a targeted approach, then it may well employ radio frequencies or microwaves. At such wavelengths the beamwidth is naturally large, and can be confined only using a very large antenna. If, however, the alien pursues a targeted approach, it would be obliged to send in the ultraviolet, visible or infrared to achieve a sufficiently narrow beam. Given how little we know of the dangers that might lurk in the galaxy, the conservative approach would be not to announce one's existence by broadcasting, especially at the powers required for communication over significant interstellar distances. In contrast, the targeted approach does not require so much transmitter power, is technologically easier, and is essentially undetectable unless someone is looking at the source at the right time and is specifically seeking a signal of that type. In targeting a particular star system, it makes sense to confine the beam to an area no greater than the habitable zone of that star. Although the beam to achieve this is very narrow for a target hundreds of light-years from the source, it is possible using an optical system with a diameter of 1 to 10 meters. To achieve the same beamwidth at radio frequencies or microwaves the transmitting antenna would have to be 1,000 times larger.

If we conclude that a short-wavelength receiver is most appropriate, then the next issue is the modulation format. It turns out that sending short, intense pulses at laser wavelengths is considerably more efficient than for a continuous-wave system. For a start, the average power of the pulse system is much less than the peak power. The average power would have to be much greater to enable us to detect a continuous-wave signal than a pulsed signal. For a short time period such as 1 nanosecond, the peak power of the laser pulse can outshine the host star and hence almost eliminate the background noise this contributes.

There remains the issue of the time the transmitter spends pointing at one star, and the time the receiver spends looking at the star which hosts the transmitter. Ignoring the time for the signal to cross interstellar space at the speed of light, the transmitter and the receiver are required to point at each other simultaneously to establish a link. Later, we will discuss a strategy which could enable us to achieve this. For now, we merely conclude that it is more likely that a sending civilization will use the targeted approach.

A civilization signaling by laser will try to make it easily discernable from noise. As the field of view of the receiving telescope will include the host star, the sender will avoid wavelengths at which the star is particularly intense. Hence, for class M stars, which are less luminous and redder than the Sun, the wavelength chosen for a signal would be different than we would use if we were doing the transmitting. Figure 2.4 illustrates the blackbody radiation energy distribution curves of different classes of star. The Sun has a surface temperature of 5,700K and the wavelength of the peak output is around 500 nanometers. However, a main sequence star of 4,000K peaks at 754 nanometers, and one of 3,000K peaks at 966 nanometers, well into the infrared. As a result, the energy in sunlight at 500 nanometers is about seven times that of a 4,000K star and an order of magnitude more than a 3,000K star. Thus a laser signal from an M class star would likely be towards the blue end of the visible spectrum to make it readily distinguishable from the host star. Although the majority of main sequence stars are class M, and their background noise would pose little problem to a wideband optical receiver, there are, as noted, reasons to believe that complex life will not develop in such star systems.

The question of humans sending signals

There may be excellent reasons for a civilization not revealing its existence. We face this dilemma ignorant of the risk. In fact, the International Academy of Astronautics has drafted a document stipulating the conditions under which nations might consent to the sending of a signal. Specifically, it sets out the following basic principles. The decision to transmit a message that might be detected by an alien intelligence should be made by an appropriate international body, broadly representative of humankind. If the decision is made to send a message, then this should be sent on behalf of all humankind, rather than individuals or groups. The content of such a message should be drawn up by an appropriate international process, reflecting a broad consensus. But at present there is no way to prevent anyone from transmitting into space. The first message sent from Earth is attributed to none other than Frank Drake. To mark the inaugural ceremony of the Arecibo telescope in 1974, he sent a message at 2.38 gigahertz that lasted for 3 minutes. It was aimed at M13, which is a globular cluster of hundreds of thousands of stars. This was selected on the assumption that the sheer number of stars would increase the likelihood of there being an intelligent society in the beam. However, as noted in an earlier chapter, a high star density would seem to preclude the long-term stable environment required for the development of complex life. The likelihood of there being an advanced civilization in this cluster is further reduced by the fact that it consists primarily of old stars which have evolved off the main sequence. However, as M13 is 25,000 light-years distant we probably did not give ourselves away by this transmission. When Drake sent messages to four nearby solar-type stars in 1999 he was criticized. Although there is much more sensitivity today to anyone deliberately transmitting to the stars, in 2006 the French sent a TV transmission to Gamma Cephei, also known as Errai, some 40 light-years away. But this is a binary star with a primary that is of spectral class K and is evolving off the main sequence, accompanied by a red dwarf, and although the primary is known to possess at least one planet the system is unlikely to host an advanced civilization.

Whilst we can transmit digital signals, it is not evident how we should transmit so that a civilization can 'read' the message. Understanding how best to do this is a complex issue. Although an alien could receive the signal and decode a sequence of pictures and words, whether it could comprehend the symbols and perform correct interpretation is open to question based on sample tests given to scientists here. One suggestion has been to send an iconic chemical language in the expectation that this would be more readily understood; after all, the laws of physics are universal. An artificial chemical encyclopedia can express chemistry and physics, and perhaps biology and medicine.

So what are the risks? It may be far-fetched, but one scary scenario is an advanced society discovers that sometime in the next 10,000 years rampant volcanism will kill advanced life on their planet, so they set out to find a planet to colonize. If they were to become aware of our existence before we had developed a capability to somehow defend ourselves then we would be at risk of invasion. Colonization might start with introducing a microbe to which they are immune and we are not, to enable them to eliminate biological competition. Once conditions were right, they would establish their own biosphere. If the aliens faced extinction on their own planet, an interstellar voyage lasting thousands of years might be an acceptable venture. Of course, even if someone decides to invade us, we are safe while they are in transit.

15 Other parts of the spectrum, decoding the data and forming pictures

We initiated our search for extraterrestrial intelligence as soon as we developed the means to do so, and it expands in pace with technological development. Although it was dispiriting not to have had an immediate result, it is likely that we will continue our efforts as long as our civilization exists. However, civilization is a fragile thing. It is possible, if unlikely, that we will be wiped out like the dinosaurs. A more likely scenario is that warfare, pandemics, pestilence or climate change causes civilization to fall into a new 'dark age' which lasts for hundreds of years. But if we are able to avoid such dangers, the search will continue, perhaps in fits and starts, until success is achieved. Other issues that may play a role in a successful SETI program include other parts of the electromagnetic spectrum, receiving stations off Earth, recovering the information in the pulse train, forming and interpreting pictures from such data, and methods that will resolve the transmitter/receiver directionality issue.

Many bits per pulse

As stated earlier, it is desirable to send short laser pulses which can readily outshine a host star, for easier detection. As the pulse rate need not be high, the average power transmitted is reasonable. A low pulse rate facilitates sending data at many bits per pulse. In a standard continuous-wave communication system, each pulse or absence of a pulse represents a single bit, with an equal chance of it being a '1' or a '0'. But if an M-ary system is used instead of a binary system then it is possible to send more than one bit per pulse. The fact that physics at laser and infrared wavelengths favors short pulses and a low duty cycle lends itself naturally to an M-ary system. Consider a system in which M intervals or slots are present. Each slot can represent a unique number. If we restrict ourselves to precisely one pulse in the M intervals of time T, then each pulse represents $\log_2$ M bits. This type of system does not have to take up the full period between pulses, but can function when only a small percentage of the maximum time between pulses is utilized, as shown in Figure 15.1. If we send F pulses per second, then the product of F and $\log_2$ M is the bit rate expressed in bits per second. Figure 15.2 relates duty cycle, intervals, and bits per pulse. Stating the bits per pulse in terms of M intervals and base 2 logarithm simply specifies that a certain number of '1' and '0' bits will convey the same information. Humans find the

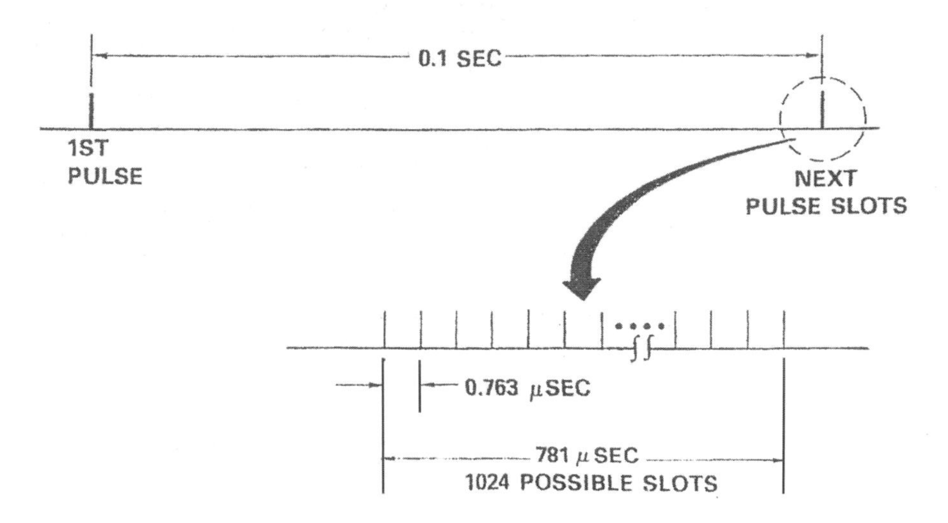

Figure 15.1. Pulse interval modulation with a 'window' showing how a small part of the interpulse can be used to convey information.

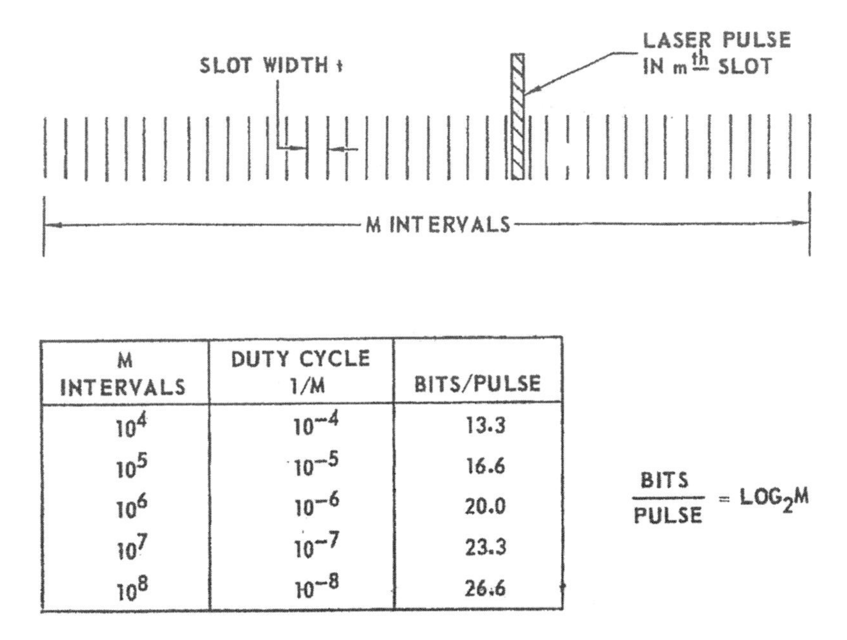

M INTERVALS	DUTY CYCLE 1/M	BITS/PULSE
10^4	10^{-4}	13.3
10^5	10^{-5}	16.6
10^6	10^{-6}	20.0
10^7	10^{-7}	23.3
10^8	10^{-8}	26.6

$$\frac{\text{BITS}}{\text{PULSE}} = \text{LOG}_2\text{M}$$

Figure 15.2. A pulse interval modulation waveform as an example of a low-duty-cycle highly efficient signal design for laser communication. The table gives bits/pulse and duty cycle in terms of M possible pulse periods or intervals.

decimal system with 0, 1, 2, 3, 4, 5, 6, 7, 8 and 9 convenient in everyday life, but every number can be converted to base 2 for use by a digital computer. A number in the range 0 to 31 can be represented by a sequence of five binary digits. The position of the '1' or '0' in the sequence determines its value. For example, in a 5-bit word the least significant bit is 2 raised to the zeroeth power. Next in line is 2 to the first power, then 2 to the second power, 2 to the third power, and 2 to the fourth power. Hence 22 in base 10 is represented as the 5-bit word 10110 in base 2, with the individual bits having values according to their positions as follows:

$$2^4 + 0 + 2^2 + 2^1 + 0$$
$$16 + 0 + 4 + 2 + 0 = 22$$

If there are M choices for the placement of a single pulse, then as M increases it is possible to carry more information in that pulse because there is greater uncertainty before the fact – just as a roulette wheel with 90 numbers provides a lower chance of the number 22 showing up than would a wheel with only 36 numbers. Information is directly related to uncertainty, in that information is conveyed only if it reduces the degree of uncertainty.

In a system which we designed, we would know M and the pulse rate. Here we do not know, but the position and rate of detected pulses can be subjected to powerful computer analysis to determine whether it is an M-ary system and then estimate the value of M and the likely pulse time slot. Here we have assumed 1 nanosecond, but computer analysis can choose different time slots and then determine the best fit of the data based on the time of detection. And as the sender will desire the signal to be interpreted, there will undoubtedly be a reference pulse to enable the recipient to establish the timing and determine M. A pulse train which repeats at some measured rate constitutes a reference pulse for another pulse whose position with respect to the reference pulse changes each time. The pulse after the reference pulse can contain a number of bits owing to the many potential time slots in which the information pulse might occur. We may not know whether the transmitter is sending using nanosecond time slots, or longer time slots, but for a short pulse of nanoseconds or less it is more likely the former, as otherwise the communication efficiency is needlessly low. It is theoretically possible to operate without a reference pulse, but this makes analysis of the data much more difficult. As an example, if there is a reference pulse once every millisecond then there will be 1,000 pulses per second, and an M of 1,000 with nanosecond time slots for the information pulse allows 10 bits per information pulse. With 1,000 pulses per second, we receive 10,000 bits per second. In such a system, the duty cycle of the transmitter remains low at 10^{-6}. This could be exploited either to limit the average power required by the transmitter, or, more probably, to enable a single transmitter to send low-duty-cycle pulses to a large number of star systems.

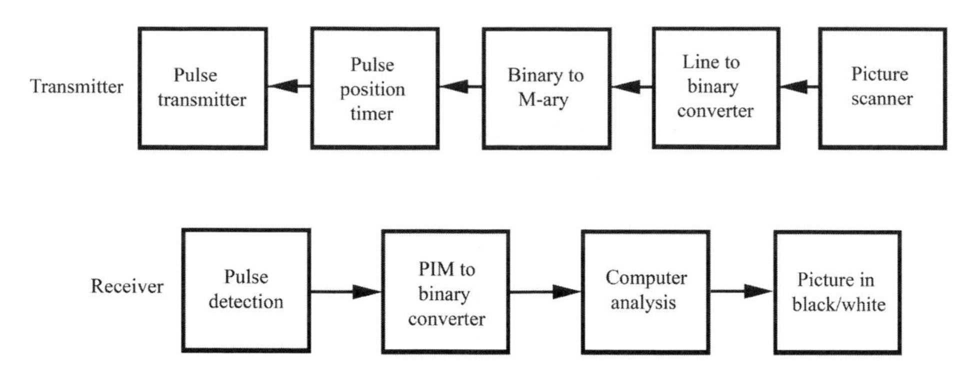

Figure 15.3. Sending pictures using low-duty pulses. Computer analysis in the receiver determines the number of pixels per line and lines per frame.

Sending and receiving pictures with pulses

A low-duty-cycle pulse system can send pictures with many bits per pulse. This can be converted into a binary number to represent a line of pixels. A succession of lines constitute a black and white picture. Figure 15.3 shows this process. At the transmitter, the picture is scanned line by line, and each line represented by a series of '1's and '0's that are converted into an M-ary number which causes the pulse to be sent at the appropriate time. For 1-nanosecond time slots, the uncertainty of 1 million choices allows 20 bits per pulse. By sending 20 reference pulses and 20 information pulses at appropriate times it is possible to send a 20 × 20 picture. Interpretation of such a simple picture might be difficult. For example, Figure 15.4 could be interpreted as showing two life forms (perhaps male and female) from the fourth planet of the star system. Greater resolution on the part of the sender would be helpful. A case of greater resolution would be if we reconstruct smaller sub-areas for a 1,000 × 1,000-pixel view. Software can examine the detected pulse train and work out if it contains a picture, determine the dimensions and can reconstruct it. Of course, the meaning of the image may elude us! One million 1-nanosecond time slots equals 1 millisecond. This is the maximum time between reference pulses, and based on 1,000 information pulses at 20 bits per pulse this puts an upper limit on the data rate of 20,000 bits per second. This could send a 1-megabit picture of 1,000 lines with 1,000 pixels per line in 50 seconds. To put this into context, 1 megabit per frame is about equivalent to a high-definition TV. This can be accomplished by a laser which operates with a duty cycle of only 2×10^{-6} by firing 2,000 pulses per second. Although black and white pictures are easy to 'see', color pictures might be very difficult to interpret because even if the data provided three color channels, we would not know the wavelengths of these colors.

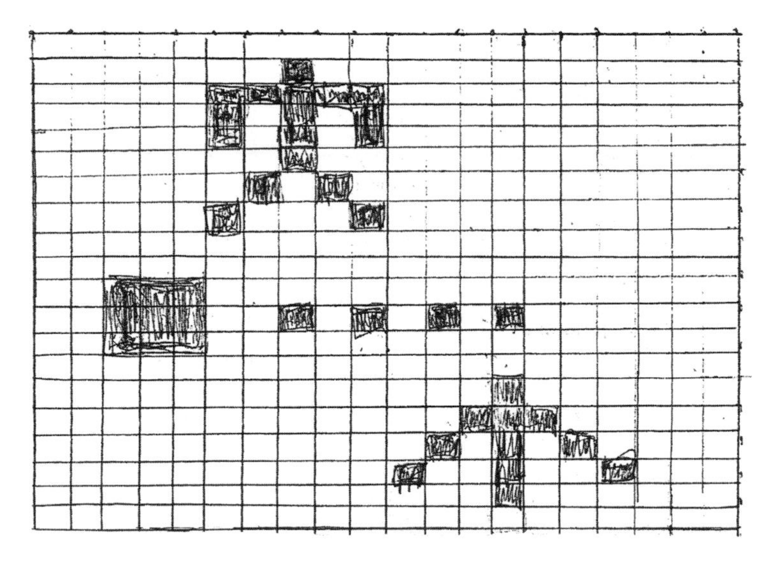

Figure 15.4. A simple picture example using 20 x 20 pixels 20 M-ary pulses at 20 bits per pulse.

Other parts of the electromagnetic spectrum

Although the microwave and optical parts of the electromagnetic spectrum appear to be the most probable choices for extraterrestrial signals, we should also consider the millimeter band, infrared and ultraviolet. Absorption by a planetary atmosphere is an issue in the choice of wavelength. In the near-infrared the predominant absorption is due to vibrational modes of the water molecule. Carbon dioxide also contributes to near-infrared absorption in bands centered at 2.7 micrometers and weak bands at 1.6 and 1.4 micrometers. Figure 15.5 shows the absorption bands in the infrared. In fact, a laser could operate in this region, avoiding the absorption bands. Solid-state lasers at 2.1 and 3.9 micrometers have been approaching a level of practicality that shows us that these wavelengths would be available to an advanced society for laser signaling.

In summary, in using ground-based receivers the visible and near-infrared are better options than the ultraviolet or mid-to-far-infrared bands that are heavily absorbed by the Earth's atmosphere.

The millimeter-wave band suffers both atmospheric absorption and noise, and does not offer the antenna gain to compete with shorter-wavelength lasers. We could probably overcome the technical issues with enough research and development, but just as microwave technology was not developed until a requirement was perceived for it during the Second World War, we will probably not make a major investment in millimeter-wave technology until there is a need for it. Useful systems currently go up to around 100 gigahertz. In addition to having few atmospheric windows, the infrared faces the problem that its

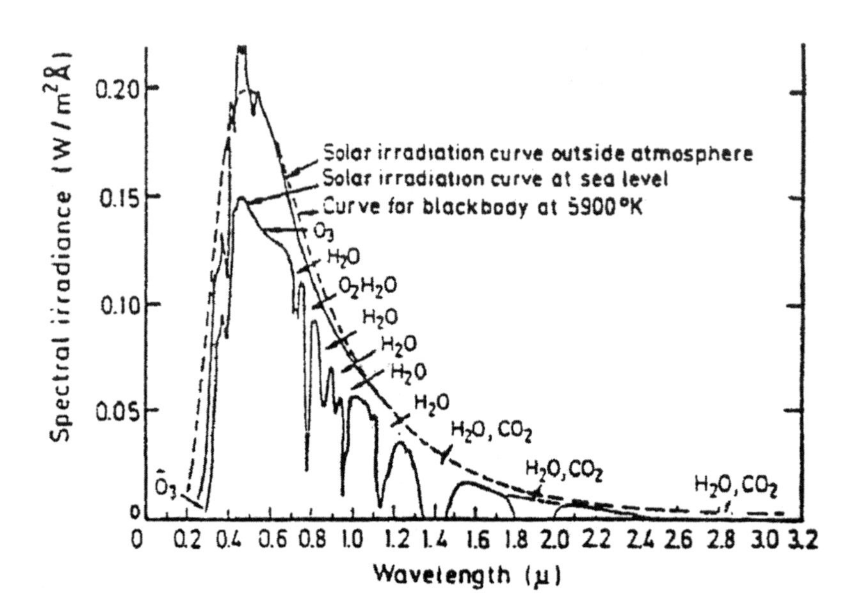

Figure 15.5. The blackbody curve for the Sun ranging between the ultraviolet and the infrared, the actual solar flux impinging on the atmosphere, and a series of atmospheric absorption bands in the infrared.

photons are not powerful enough for direct photon detection to provide sensitive noise-free devices capable of overcoming the thermal noise without the detector being cooled to around 77K. Even when cooled, infrared detectors used to suffer from significant noise owing to the lack of noiseless gain in the devices, but infrared APDs such as HgCdTe devices are now becoming available which offer high sensitivity because of their internal gain. It is still necessary to cool the detector significantly for optimum performance, however. Progress in low-noise infrared detectors in the bands below a wavelength of 4 micrometers offers another possible regime. Noiseless gain is also possible by going to a heterodyne receiver, but this requires an infrared laser local oscillator inside the receiver and this in turn obliges one either to make an assumption about the specific wavelength to detect, or to search the spectrum by tuning the local oscillator – and tuning means the receiver may miss a signal unless it is a continuous wave, because it could well be tuned to a different part of the infrared spectrum when a pulse shows up. However, with the availability of low-noise detectors with internal gain, there is little reason to take on the complexity of heterodyning. Owing to the limited receiver sensitivity for direct detection at 10.6 micrometers, a heterodyne system has been developed for infrared laser communication, but tuning it is much easier when the transmitting frequency is known! Again, when we use the term infrared here we mean from 2 micrometers to 1,000 micrometers, where it becomes the millimeter band. At shorter wavelengths, the near-infrared is more akin to the visible range because the energy of a photon

Wavelength (micrometers)	UV improvement
2.0	64
1.0	16
0.5	4

Figure 15.6. Ultraviolet lasers improve antenna gain over other wavelengths. The wavelength of UV is around 0.25 micrometers.

is sufficiently high for devices to offer almost noise-free gain, for sensitivity without resorting to complexities of heterodyning in the receiver. Heterodyning requires the incoming signal and the laser local oscillator in the receiver to be aligned with great precision so that an uncorrupted input wave can be mixed with the phase front of the local oscillator to produce a useful intermediate frequency. But if at the wavelength in question the atmosphere can seriously affect the alignment of the incoming wave then an active correction must be incorporated to eliminate phase front changes. In addition, Doppler shifts must be tracked and eliminated.

All of the above implies that infrared wavelengths longer than a few micrometers would not be a good choice for SETI searching.

Ultraviolet at a wavelength of 0.35 micrometers offers possibilities. Although a ground-based receiver is impractical due to absorption by the atmosphere, a receiver on a space station or on the Moon is attractive. The shorter wavelength of ultraviolet increases the antenna gain of the transmitter, which means the beam spreads out less in crossing interstellar space. As a result when the beam reaches us all of its power is confined into a smaller area, which means the same signal density can be obtained for a reduced transmitter power than in the case of a longer wavelength. Figure 15.6 shows the increase in antenna gain with shorter wavelengths when compared to other wavelengths. Arthur C. Clarke, who gave us the concept of the geostationary satellite and a great deal of award-winning science fiction, considered ultraviolet to be a much better option than either infrared or visible (personal correspondence with author, 1996). Since there is no air on the Moon there is no wind, so a large receiver will be able to be constructed using very lightweight materials. NASA's Goddard Space Flight Center recently reported that it had built a 3-meter-diameter dish using a concrete-like substance which consisted of crushed rock and an epoxy of carbon nanotubes. The dish was then spun in vacuum and coated with aluminum. This process could probably be scaled up to manufacture a telescope on the Moon with a diameter of 20 to 50 meters. Even if the figure of the mirror was not perfect, it would still make a practical ultraviolet 'photon bucket' for SETI. Owing to their high energy, ultraviolet photons offer high detector sensitivity. They are detected by photomultipliers of high quantum efficiency which offer noise-free gain to overcome internal noise issues. If optical SETI from the Earth's surface proves fruitless, a reasonable next step would be to operate an ultraviolet receiver in space.

Telescopes off Earth

A team of internationally renowned astronomers and opticians has proposed to make very large telescopes on the Moon using liquid mirrors. A parabolic mirror could be created using a slowly rotating ionic solution (molten salts) with an ultra-thin coat of silver no thicker than 100 nanometers. It is estimated that all the materials for a 20-meter-diameter lunar telescope would weigh only a few tons. A single Ares V rocket could boost this to the Moon in the 2020s. Future telescopes might have mirrors as large as 100 meters in diameter. They could peer back in time to when the first stars and galaxies were forming. Although such telescopes would be designed as imagers, they would certainly address the question of 'bigness' for optical collectors for SETI. That said, however, based on past history SETI would not be the observing priority for a telescope of this size on the Moon.

NASA has been considering for a while now how to unfurl antennas in space for a variety of applications at microwaves and millimeter waves. In addition, large solar arrays have been deployed to draw power from the Sun. The solar arrays that NASA installed on the International Space Station span 73 meters

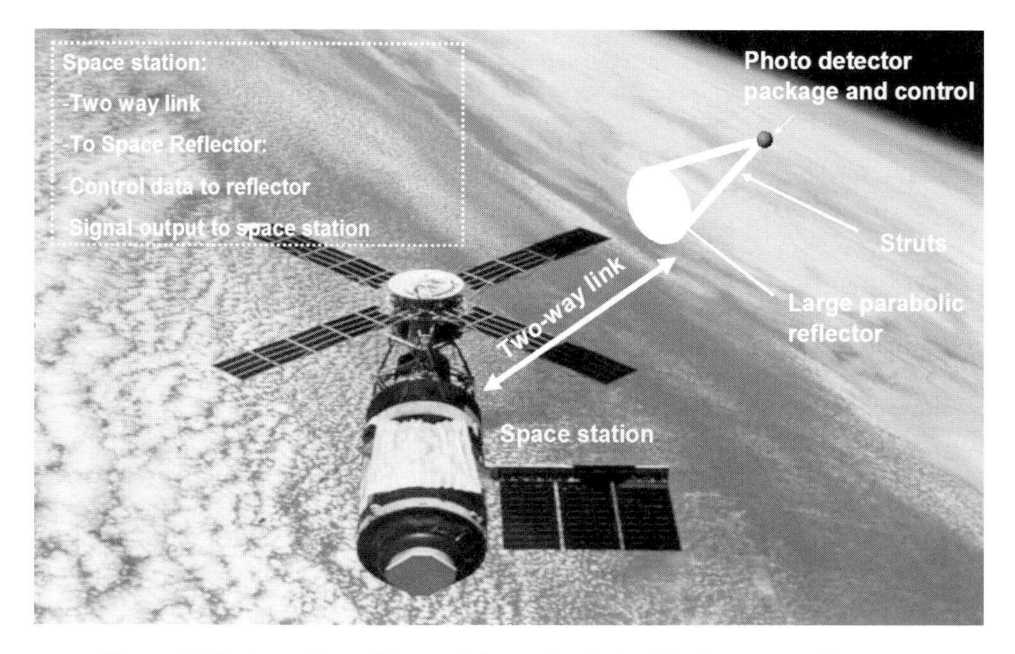

Figure 15.7. An outline of how a 'photon bucket' might be operated in space.

from tip to tip. Although the design of a solar array has some similarity to that of a large photon bucket, there are unique design challenges involved in deploying and using lightweight optics in space. In particular, the shape of the collector surface must be precisely maintained. Solar radiation and thermal distortions will require sophisticated solutions. The low residual pressure of the Earth's atmosphere at satellite altitudes induces sublimation of materials, evaporation of lubricating fluids, and vacuum welding. A stabilization system will have to overcome instability effects of torques created by solar pressure acting on large surfaces. Figure 15.7 illustrates a photon collector in space with control accomplished by a link to either a space station or some other facility away from the collector.

After spending almost half a century seeking SETI signals at radio frequencies, we are expanding the search to the optical spectrum. After a few decades we might come to the conclusion that it is preferable to build large optical collectors either in space or on the Moon. Perhaps an advanced civilization will signal using ultraviolet lasers precisely because requiring the recipient to be capable of operating in space is seen as a mark of its worthiness.

16 Future systems for intercepting alien lasers

In order to design a system that solves the directionality problem, we must estimate the number of star systems we would examine over some reasonable time period. To arrive at an estimate, we must first decide the maximum radius of our search volume in light-years (for this first go-around at least). If we are dealing with distances up to 1,000 light-years, using estimates by Margaret Turnbull and others and the many constraining factors discussed by Frank Drake and Stephen Webb, we can expect at most 40,000 likely candidates of the approximately 8 million stars within this radius. The critical issue is the time required to cycle through the list.

The same rationale applies to a civilization sending out probing signals as applies to anyone watching for signals. Let us assume that a transmitter works through about 10,000 stars per year, sending laser pulses to each. If it sends nanosecond pulses in every millisecond, then by operating at a very low duty cycle in order to provide a large peak power it can send at least 1,000 pulses to 1,000 stars in 1 second, and by using an M-ary signal format each pulse can contain 10 bits. Our task is to design a receiver which is essentially guaranteed to spot such a signal. An analogy that may be helpful is to imagine a pitcher throwing, in turn, balls to ten potential locations at which there may be catchers. But each catcher only knows a ball may be thrown at him from any of a number of directions. Yet the catcher must be facing the right direction when the ball is thrown to him. He must stop, stare for some length of time at one of the possible directions from which the ball might come, and if it has not been thrown he must turn to another possible point of origin. The length of time that he can look in a given direction is the maximum interval between pitches in terms of his context. The pitcher simply keeps throwing and repeating the cycle. He throws a pitch, then moves to another direction, doing so for ten directions before repeating. The catcher must estimate the time between repeats from any given direction, then look in each direction for at least that duration. Even if nine of ten times the pitcher is throwing where there is no catcher, and the catcher is looking in many directions and there is no pitcher in almost all directions, as long as the pitcher is in one of the directions when the catcher is looking, the ball will be caught if the catcher holds his attention long enough to include the total time between pitches before repeating. The fact that the pitcher makes all but one of his pitches to no catcher means that he must pitch at a much faster rate than the catcher changes direction, but the pitcher cannot know which pitch is being caught and he is required only to continues throwing in each of the ten directions, repeating the cycle. The key point is that the cycle time is sufficiently fast to allow the catcher to be looking at the pitcher at a time when a ball is being thrown in his direction.

A laser can transmit 1-nanosecond pulses in an interval of 100 microseconds and cover 10,000 stars every second. The rapid steering required is plausible, since it is only a small step beyond our capability today. This system places the most advanced technology in the transmitter, thus allowing the receiver to be less challenging. The receiver will inevitably run at a much slower rate. It might look for 1-nanosecond pulses for a second or so, and then slew to the next candidate in a few milliseconds. It could continue doing this until every star on the list has been examined, or it could give a subset of stars a longer look before advancing to another subset. The strategy given in this chapter solves the directionality problem. With laser signals occurring every millisecond and the receiver looking for a second, when the alignment is right the receiver must see the laser pulse. The key point is that a detection will still occur even if the pulse width, pulse rate, and number of slots in an M-ary system are off the mark. The transmitter frequency need not be known, since for this 'double-scan' approach only the general band of the spectrum (visible, ultraviolet, infrared) needs to be correct. If the transmitter can probe 10,000 stars in a year and there are 40,000 stars on the candidate list, then the 'revisit time' ought to be no longer than 4 or 5 years. This is an insignificant fraction of the light travel-time between stars in the galaxy.

Search strategy summary

The logic of the strategy is based on the following assumptions. Since there are too many candidates, the transmitter will not be dedicated to a single star system. Hence, it will work through a list, briefly sending to one candidate before rapidly slewing to the next. The time that the receiver spends looking at a star must include at least one night-time period on that planet, hence a key factor will be the longest likely rotation time of a habitable planet. As the Earth's rotation period is 24 hours, a sender might consider at least twice and perhaps four times this. Assume 100 hours as the limit. If the limits of transmitter antenna size and accuracy of pointing permit the minimum useful beamwidth to be adjusted for each candidate star, then interstellar distance is not an issue: a 1-microradian beam will be sufficient at a range of 50 light-years, and a 0.1-microradian beam at 500 light-years. It is therefore entirely feasible to deliver all of the power within the habitable zone of a star out to a range of 500 light-years. The search strategy involves various times of interest arranged to optimize detection by the receiver and to avoid missing signals. Figure 16.1 shows the key system timing parameters of one viable configuration. Figure 16.2 shows the concept of the scanning transmitter. Figure 16.3 is divided into two parts to show the timing sequences: one part detailing the transmitter requirements and the other the receiver requirements. In this configuration, we use an M-ary system that allows 10 bits per pulse and places some transmitter data rate limits. It is

	Value	Reason
Pulse Detection System	1ns	Short pulse low duty cycle likely
Time looking continuously at one star system	<10 sec.	Assumes pulses will not be further apart
Time between pulses	~1 microsecond	Enable 1,000 ~1Ns. Slots for 10 bits/pulse
Duty Cycle	<1%	Enable high peak power while constraining average power
Time including same star system	<100 hrs	Based on likelihood of message length if based on likely rotation rate of habitable planets

Figure 16.1. Key system timing parameters.

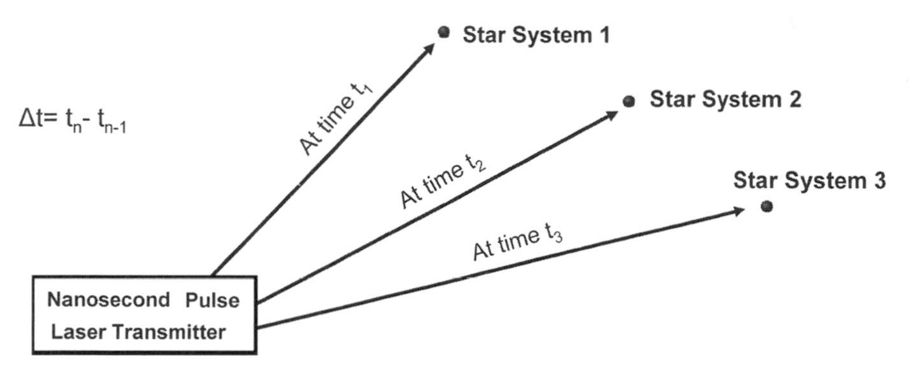

Figure 16.2. A laser transmitter rapidly scanning star systems.

possible to lower the bits per pulse and increase the number of stars addressed in one sequence. For example, if we sent a binary system of M=2 (e.g. 'on' or 'off') then we can send a pulse every 100 nanoseconds and send to 10,000 stars in one sequence at a rate of 10,000 pulses during the 1 millisecond of time during which the beam is on a star. The transmitter moves to another candidate

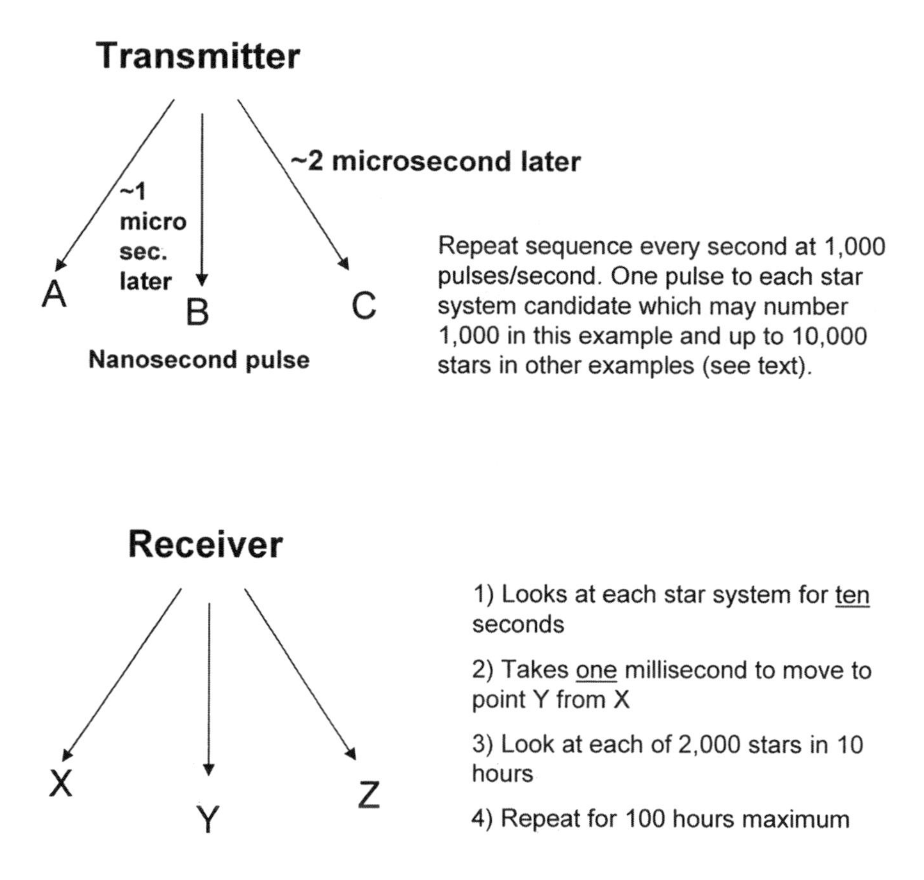

Figure 16.3. Timing sequences.

every 100 nanoseconds, 10,000 times per second. Hence, because M=2, the bits per pulse is 1 and there will be 10,000 bits per second achieved. Because 10,000 pulses at 1 bit per second replaces the earlier example of 1,000 pulses per second at 10 bits per pulse, the duty cycle of the transmitter will increase by a factor of 10, but is still only about 1 per cent. The exact duty cycle is dependent on the specific pulse width and the average pulse rate.

First, let us examine the transmitter requirements in a low-duty-cycle system. We assume that the laser pulse is short, say 1 nanosecond, to overcome the background light of the host star. The time between pulses is a maximum of 10 seconds. It is likely to be shorter, but the strategy should be able to accommodate pulses that far apart. The message to each star should be repeated at least twice to allow for the possibility that a receiver missed either the beginning of the message or the end of the message. The time between revisits is arbitrary, and is dependent on the number of candidates the transmitting society considers worthy of attention, and how many transmitters it has. It seems reasonable that the time between transmitter revisits will be a small fraction of the signal's travel-

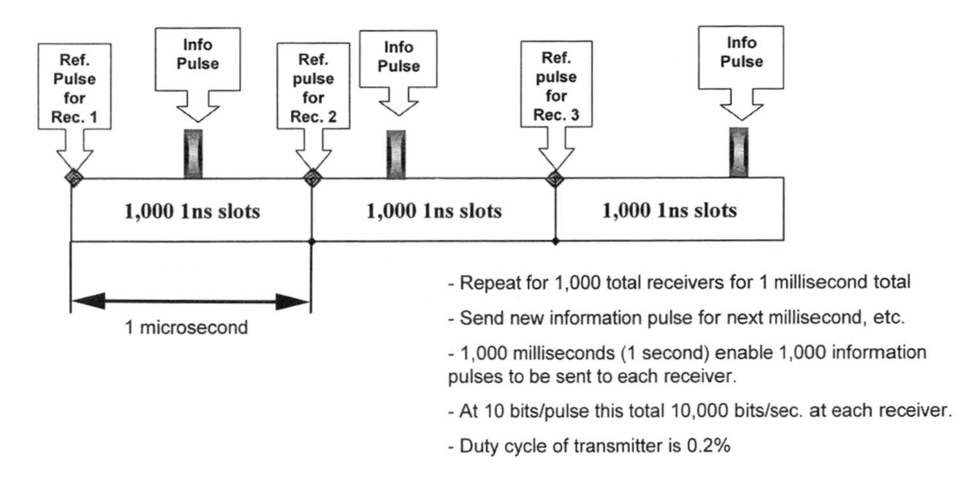

Figure 16.4. Timeline of a scanning transmitter.

time. There would be little cause to return more frequently than once per year. Say 5 years. One can speculate that if the lifetime of the senders is much longer than for a human, then the revisits may be much further apart. We are describing a system with a very low laser duty cycle. With 1 second maximum on a given candidate, then this is a duty cycle of 10^{-9} as seen by one star system receiver. This enables the transmitter to send pulses to other candidates in the time between: 1,000 candidates can be addressed without raising the duty cycle of the transmitter above 10^{-6}. Figure 16.4 shows a timeline of the scanning transmitter for 10 kilobits per second to 1,000 candidates at a duty cycle of 0.2 per cent. This scenario assumes the sender has cut its list down to 1,000 candidates. Despite our inferior knowledge, we can still conduct a valid search for a signal with a list of 40,000 candidates.

How should we design a receiver to successfully detect this source? The issue of a receiver's size was discussed in an earlier chapter. It depends primarily on what the transmitting society thinks is a serious effort at reception, and what they think of as being 'big'. Given that we make the receiver large enough (however large this turns out to be), how do we search to maximize the probability of detection? At some time during the search, both the transmitter and the receiver must be looking at each other simultaneously (ignoring signal travel time). If the receiver were to examine a total of 40,000 candidates within a radius of 1,000 light-years, it could spend a significant time on each star. If the receiver looked for 10 seconds at a star and then moved in milliseconds to the next star, it could cover six stars per minute. It could cover 3,600 stars during a 10-hour night. This assumes essentially no switching time to slew the receiver between stars. If we allow several seconds for each movement, then it is still possible to examine at least 3,000 stars per night. The receiver can look for no more than about 100 hours before moving onto another subset, and it will not return for a while. As our day is only 24 hours, we can fit four days and nights into a 100-hour period, and looking for more than 100 hours is unlikely to be productive. As noted, the more we learn about the planetary systems that exist around other stars, and the better we understand the constraints placed on the development of intelligent life in such systems, the more we will be able to trim our list of candidates, simplifying our search.

The message that the aliens send should not exceed 10 hours in duration before repeating, since a longer message could easily be interrupted by a period of daylight on the recipient's planet. If we assume 10 hours is the minimum period of planetary rotation consistent with the development of intelligent life, the message must be no longer than 3 hours in length to ensure that it can be received in darkness, and should be sent at least twice, with a suitable interval in between, to ensure that one complete transmission is able to be received.

If our selection of candidates is correct, and there is someone signaling, then this strategy will succeed in a reasonable time, by which we mean within a decade.

Data rates

Can large data rates be obtained with the general approach outlined? We can assume a pulse interval modulation (PIM) system of 10 bits per pulse, with only one pulse per second for 3 hours of transmission resulting in a message length of the order of 100,000 bits. The message is then repeated. This is likely to be at the low end of the range of possible data rates. A higher pulse rate will increase the data capacity at the expense of either raising the duty cycle (which is very low, and so could bear an increase) or reducing the number of stars sent to. We can presume that the sending civilization will know more about space science than we do, by which we mean they will have a shorter list. The pulse rate can readily go to 100 per second, yielding a message length of about 10 million bits. A rate of 1,000 pulses per second will be a duty cycle of 10^{-6}, which is still low, and allow a message length of 100 million bits. This amount of data is sufficient to be useful, including stating what size to build an antenna for a higher rate information channel.

Once we have positively detected a signal, we would soon find the funds to build the follow-up receiver! Whether it consisted of a single large collector or an array of smaller ones, a massive effort would be awarded priority status.

The ability of a laser to point to one star, send an intense short pulse, then slew to the next star, and so on in succession, would seem to be well within the grasp of an advanced civilization. A single large pulsed laser could target thousands of stars in a relatively short time, as measured in Earth years. It could send hundreds and perhaps thousands of pictures to each of the stars believed likely to host intelligent life. This can all be accomplished using short pulses where each pulse carries a number of bits indicated by the time slot in which it occurred between reference pulses.

The ability of a listening civilization to look at all the likely candidates renders moot the argument that a broadcast mode is necessary on the basis that one simply doesn't know where to point.

Pictures in the data stream

Using 1 megabit per picture gives a high resolution, and at 10,000 bits per second it will take 100 seconds to receive a picture, which is potentially 360 pictures during a 10-hour night of receiving. Meanwhile, owing to the low duty cycle, the laser will also be sending to other stars. At 1,000 time slots per nanosecond pulse and very fast slewing, a star will receive 1 microsecond of attention before the beam slews, which is 1,000 stars per millisecond. Then the process repeats. Since one pulse to each star occurs in a total of 1 millisecond, 1,000 such cycles per second send 10,000 bits per second. Figures 16.3 and 16.4 show sending one pulse to one star system followed by a pulse to another star, etc, with the information pulse being sent at the appropriate time. The duty cycle of 2,000 pulses, each of 1 nanosecond, sent to 1,000 stars is an overall duty cycle of only 0.2 per cent, meaning that the laser is off for over 99 per cent of the time. If we change the parameters a bit, we can, as noted earlier, increase the data rate. Low-duty-cycle systems can still yield high pulse rates. For example, we can assume 100 picosecond pulses and time slots (10 per cent of a nanosecond). This allows the pulse rate to be increased to 20,000 pulses per second and maintain 10 bits per pulse, thereby enabling 10,000 pictures to be sent over a period of 3 days. Even with this signal being sent to each targeted star, it is still only a 2 per cent duty cycle at the transmitter.

As the receiver field of view is small, the solid angle, Ω, is approximately equal to the square of the simple angle, θ, in steradians. If θ is about 1 milliradian, then θ^2 is 10^{-6} steradians. The change of angle to point at another candidate can be quite small. On average, the more stars there are to examine, the smaller the angle between stars through which the receiver's field of view must move. Given our own technology and improvements likely to have been made by an advanced society, it is reasonable to expect fast electronic steering over such angles. A key feature of this strategy is how it makes the receiver simple. In particular, the receiver's field of view does not require to be as narrow as that of the transmitter. The transmitter has to perform the difficult task of aiming its beam with sufficient precision to deliver its energy in the habitable zone of the target star, and to slew between targets much more rapidly than the motion expected of the receiver. The transmitter moves in angle at high speed (picoseconds) and with extreme precision (nanoradians).

This dual-scanning satisfies the requirement that the receiver point at a star at the same time as a transmitter at that star is sending in the direction of the receiver. The examples above also illustrate why it is unlikely any alien signals will be found until a system like that described is available. We have the technology, we lack only the money. The will to build this system can only come

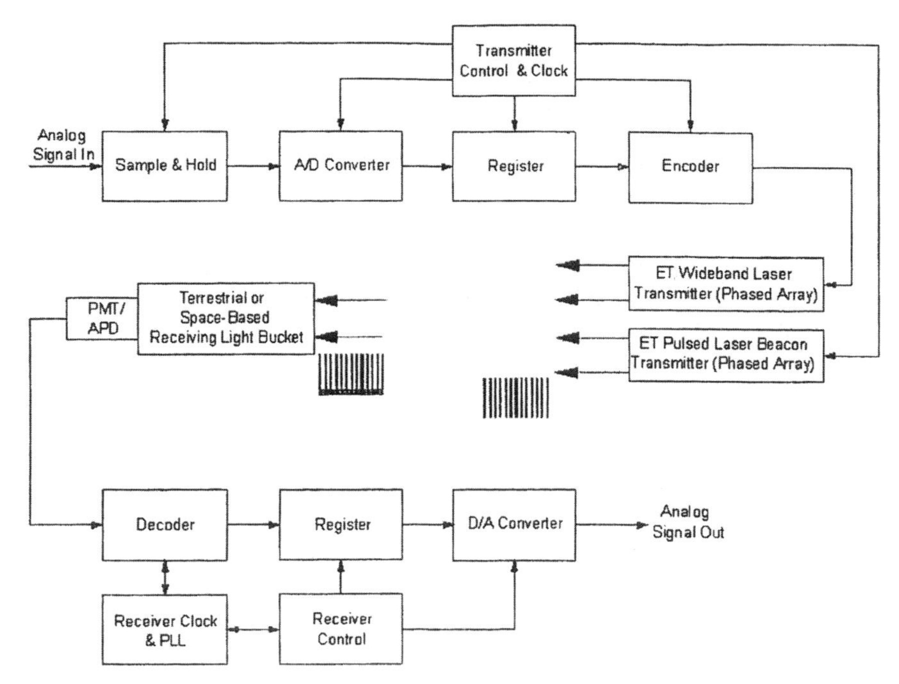

Figure 16.5. Block diagram for beacon and high-rate information channels.

about after the realization that it is a systems engineering problem that requires a systems approach to solve all of the issues. Unfortunately, this is unlikely to occur until we grow tired of looking in other ways without a positive result.

Beacon and high-data-rate channels

There is the question of whether we will first receive a beacon that will lead us to a channel where a reasonable data rate will provide detailed information. It is possible to receive a beacon on the same channel as an information channel; one just needs to build a much larger collector, because the signal per pulse is much less owing to the greater number of pulses sent per second to convey the information. Figure 16.5 shows that the same block diagram can apply. At the transmitter, a separate laser source at the same wavelength is utilized with a reduced peak power and a much greater pulse rate. The larger collector permits detection of a weaker signal pulse as the pulse rate of the channel increases, allowing greater information flow. A major advantage of combining the beacon and the information channel is that the beacon needs only to make its existence evident; it does not need to send much, if any, data. Perhaps, the only data the beacon needs to convey is how big the receiver should be to receive the information. A very low repetitive pulse rate can be used where pulses are either there or not, conveying ‘1’s and ‘0’s.

Optimum design for an optical SETI receiver

- Large effective collector area
- Multi-pixel sensor
- Single photon nanosecond response low-noise detector
- Nanosecond electronics
- Massive computer power for real-time processing

Let us examine each of the above. First, by using a really large collection area we ensure that even if the power density of the signal is deliberately low we will be able to distinguish it from the noise and detect it. At present, there is no optical collector in use for SETI that comes close to meeting this ability. To obtain the requisite high sensitivity, especially if the power density is low, we need to detect a signal with a single photon. We need fast electronics in order to achieve the pulse resolution that is needed to eliminate data errors. Massive computer power in real-time is necessary to deal with all the inputs to the sensor, to recognize potential false detections and to distinguish a valid signal. There is a complex tradeoff for the optimum number of pixels for the sensor, since it depends on the collector area and hence on how much background light there will be in the field of view of each pixel, and the likely false alarms that will have to be processed.

Multi-pixel sensor

To scan more of the sky in a shorter time, it is possible to use a multi-pixel sensor as the detector in optical and near-infrared receivers. The Harvard all-sky system uses a multi-pixel sensor. As the optics are pointed at the sky, each pixel sees a different field of view. Hence each pixel of the sensor will respond only to a signal emanating from a small section of the overall field of view of the sensor. If a sensor has a 3×3 detector array and each pixel is represented in the far field by 1 milliradian, then the sensor has an 'instantaneous total field of view' of three milliradians on each side. Semiconductor photodetector arrays are available for various applications as large as 128×128 pixels and are made using materials suitable for that wavelength regime. There are a number of issues to deal with if one is to employ a multi-pixel array. For example, it is easier to obtain high sensitivity by using a single detector rather than a number of detectors (In the former, we need a beam splitter and two photodetectors in order to eliminate internal noise pulses). Also, the Geiger mode of an Avalanche Photodiode Array (APD) can be triggered by a single photon, and its high quantum efficiency makes it a sensitive sensor. But an array of these is quite difficult to build due to uniformity issues whereby one pixel may have significantly lower sensitivity or be much noisier than its adjacent pixel. APDs have a higher quantum efficiency than photomultipliers at wavelengths in the near-infrared where crystal lasers such as Nd:YAG are high powered and efficient. Photomultipliers are more sensitive in the visible regime. But the fact that they are much larger limits the useful size of a photomultiplier array. The electronics to process the data from a multi-pixel array is considerable, since the system is traveling and not locked on a star. Thus, a different pixel an instant later will be looking at the same spot in the sky that an adjacent pixel just observed. With many pixels, a lot of fast processing is required to ensure a genuine signal is not overlooked. The amount of background light will also be influenced by the system's optical filter. That is, the fraction of the spectrum the sensor is looking at will define how many photons per nanosecond will be detected from a target star of a given spectral class. When the signal wavelength is unknown, one would want as wide a spectral range at the input as possible. It is a virtue of optical SETI using direct detection that the wavelength of the signal need not be known, only its general band (e.g. visual or near-infrared). Another assumption is that the instantaneous field of view of each pixel will receive background from only one star. This puts another restriction on the total instantaneous field of the sensor, as complexity and difficulty rise rapidly with the number of pixels in the sensor.

Summing up the double-scan system

As we have shown, an advanced civilization could readily send signals to thousands of stars by using only a few transmitters issuing short laser pulses in narrow beams, and an eager recipient would need only a few receivers to reliably search thousands of stars for such a signal. (There must be at least one receiver for each hemisphere for full sky coverage.) By trading the time on any given star for a greater number of targets, one laser and one receiver can handle thousands of targeted stars. By this, we mean that if one can point quickly and accurately, the pulses are so narrow that many targets can be achieved and still retain a low duty cycle for the transmitter. For the link to be achieved, the transmitter and the receiver must be pointed at each other at the same time, a condition that this double-scan system can achieve.

In conclusion

This book was written as the Allen Telescope Array to detect microwave signals and the Harvard all-sky system for laser signals were commissioned. These are the most ambitious enterprises of their types to-date. One of them might detect a signal. What if they fail? We will face either the Fermi Paradox that we are alone in the galaxy, or the argument that our collectors are still too small. If we have to make yet another jump in capability, we will face the decision of where best to invest our resources. A key issue will be the choice of wavelength. Let us end by reviewing this contentious issue.

Let us say that you are an alien on a planet around another star, and you desire to contact intelligent beings in some other system. There is near-zero likelihood of two civilizations being at the same level of technology. Mankind has had radio for about 100 years and lasers for less than 50 years. Let us say that your technology is several hundred years ahead of humans. You can only make yourself known to a civilization whose technology would enable them to detect your signal. Are you going to build a 1,000-meter-diameter radio-frequency antenna, or a 1-meter laser transmitter? Are you going to make one that slews slowly from star to star, or one that can do so very rapidly? Are you going to choose an approach where the recipient needs to know the precise frequency, or one that can be detected using a broadband receiver that looks at a significant portion of the electromagnetic spectrum? Are you going to choose a system that can search many hundreds and possibly thousands of light-years away? Are you going to avoid broadcasting the fact that you exist? How will you protect yourself while actively searching for an alien civilization? Will you start off with the most appropriate technology for point-to-point communication once you have found someone to talk to? Considered from this perspective, it is probably an Earth-centric delusion for us to seek a radio signal that is deliberately aimed at us and maintained continuously. But even if you chose to use a laser, perhaps you would not use one in the portions of the electromagnetic spectrum we have been considering. Perhaps you would prefer an X-ray laser. If not a laser, then it might be something we have yet to discover. But we, as the potential recipient, must choose where to make our main effort. Half a century of listening at radio frequencies has been fruitless.

We have only just begun to search for laser signals. In a sense the Harvard system is only the opening gambit. We may well have to undertake a larger project before we have a real chance of detecting a signal, if one is there. We should apply our best intelligence to the problem. To search in a comprehensive way requires very little of our society's resources. We cannot conclude that we are alone until we have made a satisfactory search, and even in that situation we should ask ourselves what else we could and should have done. To conclude that we are alone would lead nowhere. To find other intelligent beings around other stars would be as momentous a discovery as is it possible to make.

Appendix

Key space and physics numbers:

1 light-year = 9.46 $\times$ 10^{15} meters
1 light-year = 63.24 $\times$ 10^{3} AU
1 parsec = 30.85 $\times$ 10^{12} kilometers
1 parsec = 206,265 AU
1 parsec = 3.26 light-years
1 AU = 150 $\times$ 10^{11} meters
1 AU = 4.85 $\times$ 10^{-6} parsecs
Planck's constant 'h' = 6.6 $\times$ 10^{-34} joule-seconds
Boltzmann's constant 'k' = 1.4 $\times$ 10^{-23} joules per degree on the Kelvin scale
1 nanometer = 10 angstroms
1 degree in angle = 17.5 milliradians
1 arc-sec = 4.88 microradians ($\sim$5 microradians)
1 hertz = 1 cycle per second

Bibliography

Ashpole E., *The Search for Extraterrestrial Intelligence,* Blandford, 1989
Barrow, J., *The Artful Universe,* Penguin, 1995
Black, H.S., *Modulation Theory*, Van Nostrand, 1953
Boss, A., *The Crowded Universe,* Basic Books, 2009
Bracewell, R.N., 'Communications from Superior Galactic Communities', *Nature,* 186: 670, 1960
Bracewell R.N., 'Radio Signals from other planets', *Proceedings. Institute Radio Engineers*, 50: 214, 1962
Brown, R.H, and Lovell, A.C.B., *Exploration of Space by Radio,* Wiley, 1958
Calvin, W.H., *How Brains Work*, Phoenix, 1996
Cameron, A.G.W. (ed), *Interstellar Communications,* W.A. Benjamin, 1963
Clark, S., *Extrasolar Planets,* Wiley–Praxis, 1998
Clark, S., *Towards the Edge of the Universe,* Springer–Praxis, 1999
Cocconi, G., and Morrison P., 'Searching for interstellar communications', *Nature,* 184: 844, 1959
Davies, P., *Are We Alone?,* Penguin, 1995
De Villiers, Marq, *The End: Natural Disasters, Manmade Catastrophes, and the Future of Human Survival,* St. Martins Press, N.Y., 2008
Dole, S.H., *Habitable Planets for Man,* Blaisdell, 1964
Drake, F., and Sobel D., *Is Anyone Out There?,* Dell, 1992
Fisher, David, *Fire and Ice: The Greenhouse Effect,* Harper & Row, 1990
Hart, M.H., 'Habitable Zones around Main Sequence Stars', *Icarus,* 33: 23–39, 1979
Hart, M.H., 'An Explanation for the Absence of Extraterrestrials on Earth', *Quarterly Journal of the Royal Astronomical Society,* 16: 128–135, 1975
Hart, M.H., and Zuckerman, B. (eds.), *Extraterrestrials: Where Are They?,* Cambridge, 1995
Heidmann, J., *Extraterrestrial Intelligence,* Cambridge, 1986
Horowitz, P., Coldwell, C., et al, 'Targeted and All-Sky Search for Nanosecond Optical Pulses at Harvard-Smithsonian', *Proceedings of the Search for Extraterrestrial Intelligence in the Optical Spectrum,* Kingsley, S. (ed), SPIE - The International Society for Optical Engineering, 2001
Howard, A., et al, 'All-Sky Optical SETI', *Proceedings 54th International Astronautical Congress,* Bremen, Germany, 2003
Howard, A., et al, 'Initial Results from Harvard All-sky Optical SETI', *Proceedings 57th International Astronautical Congress,* Valencia, Spain, 2006
Huang, Su-Shu, 'The limiting sizes of Planets', Publications of the Astronomical Society of the Pacific, 72: 489,1960

Huang, Su-Shu, 'Occurrence of Life in the Universe', Americal Scientist, 47: 397, 1959
Jakosky, B., *The Search for Life on Other Planets*, Cambridge, 1998
Jones, B.W., *The Search for Life Continued*, Springer–Praxis, 2008
Kasting, J., Whitmire, D., and Reynolds, R. 'Habitable Zones around Main Sequence Stars', *Icarus* 101: 108–128, 1993
Kingsley, S. (ed), *Search for Extraterrestrial Intelligence(SETI) in the Optical Spectrum*, Proceedings SPIE - The International Society for Optical Engineering, vol. 1867, 1993
Kingsley, S., and Lemarchand, G., *Search for Extraterrestrial Intelligence (SETI) in the Optical Spectrum II,* Proceedings SPIE - The International Society for Optical Engineering, vol. 2704, 1996
Kingsley, S., and Bhathal, R. (eds.), *The Search for Extraterrestrial Intelligence in the Optical Spectrum III.* Proceedings SPIE - The International Society for Optical Engineering, vol. 4273, 2001
Kurzweil, R., *The Singularity is Near: When Humans Transcend Biology*, Penguin, 2005
Lemarchand, G. and Meech, K. (eds), *Biosatronomy '99, A New Era in Bioastronomy*, Publications Astronomical Society of the Pacific, vol. 213, 2000
Macdonald, D.K.C., *Noise and Fluctuations*, Wiley, 1962
Mason, J. (ed), *Exoplanets: Detection, Formations, Properties and Habitability*, Springer–Praxis, 2008
Naeye R., 'SETI at the Crossroads', Sky & Telescope, November 1992
Oliver, B.M., 'Some Potentialities of Optical Masers', Proceedings Institute Radio Engineers, 50: 135, 1962
Rodda, S., *Photo-Electric Multipliers*, Macdonald & Co., 1953
Rose, A., 'The Sensitivity of the Human Eye', *Journal of Optical Science of America*, vol.38 no. 2, February 1948
Ross, M., 'Search laser receivers for interstellar communications', Proceedings IEEE, vol. 53, no. 11, November 1965
Ross, M., *Laser Receivers*, Wiley, 1966
Ross, M. (ed), *Laser Applications*, vol. 1, Academic Press, 1971
Ross, M., 'New Search for Extraterrestrial Intelligence', IEEE Spectrum, November 2006
Schwartz, R.N., and Townes, C.H., 'Interstellar and Interplanetary Communication by Optical Maser', *Nature*, 190: 205, 1961
Tough, A. (ed), *When SETI Succeeds: The Impact of High-Information Contact,* Foundation for the Future, 2001
Von Hoerner, S., 'Search for Signals from other Civilizations', *Science*, 134: 1839, 1961
Ward, P., and Brownlee, D.E., *Rare Earth: Why Complex Life Is Uncommon in the Universe*, Springer–Copernicus Books, 2000
Webb. S., *If the Universe is Teeming with Aliens... Where is Everybody?*, Springer–Copernicus, 2002
Zebrowski, E., *Perils of a Restless Planet*, Cambridge University Press, Cambridge, 1999

Index

Printing: Mercedes-Druck, Berlin
Binding: Stein + Lehmann, Berlin